Burning Up

A Global History of Fossil Fuel Consumption

Simon Pirani

First published 2018 by Pluto Press
345 Archway Road, London N6 5AA

www.plutobooks.com

Copyright © Simon Pirani 2018

The right of Simon Pirani to be identified as the author of this work has been asserted by him in accordance with the Copyright, Designs and Patents Act 1988.

British Library Cataloguing in Publication Data
A catalogue record for this book is available from the British Library

ISBN   978 0 7453 3562 9   Hardback
ISBN   978 0 7453 3561 2   Paperback
ISBN   978 1 7868 0312 2   PDF eBook
ISBN   978 1 7868 0314 6   Kindle eBook
ISBN   978 1 7868 0313 9   EPUB eBook

This book is printed on paper suitable for recycling and made from fully managed and sustained forest sources. Logging, pulping and manufacturing processes are expected to conform to the environmental standards of the country of origin.

Typeset by Stanford DTP Services, Northampton, England

Simultaneously printed in the United Kingdom and United States of America

Burning Up

This book is dedicated to the memory of my friend Pavel Sheremet, who was assassinated on 20 July 2016 in Kyiv, Ukraine. The fearless way that Pavel worked, as a journalist in Belarus, Russia and Ukraine, was and is a profound inspiration.

Contents

Figures viii
Tables ix
Acknowledgments x
Units of measurement xi
Acronyms and abbreviations xiii

Introduction 1

PART I: CONTEXTS

1. Fossil fuels before 1950 9
2. Energy technologies 25
3. Energy in society 38
4. Fossil fuel consumption in numbers 53

PART II: CHRONOLOGIES

5. The 1950s and 1960s: post-war boom 79
6. The 1970s: crises and oil price shocks 93
7. Patterns of electrification 107
8. The 1980s: recession and recovery 122
9. The 1990s: shunning the global warming challenge 138
10. The 2000s: acceleration renewed 153

PART III: REFLECTIONS

11. Interpretations and ideologies 173
12. Possibilities 181
13. Conclusions 193

APPENDICES

Appendix 1. Measuring environmental impacts, energy flows and inequalities 201
Appendix 2. Additional figures and tables 208

Notes 212
Further reading and bibliography 247
Index 249

Figures

1. Fossil fuel production, 1800–2009   10
2. Fossil fuel production in the first and second Industrial Revolutions   15
3. Energy conversion steps   27
4. Commercial fossil fuel consumption and population, 1980–2015   48
5. Energy consumption per person per year, 1971–2011   51
6. Consumption of commercially traded energy, 1965–2014   56
7. Nigeria: domestic final use of energy, and crude oil supply   66
8. Global fossil fuel consumption, 1965–2015. Amounts by fuel, and fossil energy as a proportion of total commercial energy consumption   154
9. Commercial fossil fuel consumption in selected countries, 1980–2015   156
10. Commercial energy consumption per person per year, 1925–68   208

Tables

1.	Units of measurement of energy	xi
2.	Commercial energy consumption by major world regions and by source, 1913–68	19
3.	People without access to electricity, and people relying on biomass for cooking	42
4.	Fossil fuel consumption, 1950–2015: overview	55
5-a.	Global primary energy supply, and transformation of energy	59
5-b.	Global final consumption of energy	60
6-a.	USA: total primary energy supply, and transformation of energy	62
6-b.	USA: final consumption of energy	63
7-a.	China: total primary energy supply, and transformation of energy	64
7-b.	China: final consumption of energy	65
8.	Commercial energy consumption through two oil shocks	97
9.	Oil consumption in selected countries, 1965–85	98
10.	China's role in the 2000s	157
11.	Nigeria energy balance, and uses of crude oil	209
12.	Global uses of electricity	210
13.	Shares of global oil production	210
14.	Prices of petrol and electricity, 1980	210
15.	Industry's role in fossil fuel consumption	211

Acknowledgments

Many people have helped me to write this book. I am greatly indebted to Kolya Abramsky, Lucy Baker, Patrick Devine, Doug Koplow, Greg Muttitt and Steve Smith, who read and commented on the manuscript, and to Philip Edwards, for help with the mathematics touched on in Appendix 1. At the Oxford Institute for Energy Studies, my colleagues Bassam Fattouh, James Henderson, Anouk Honore, Malcolm Keay, Howard Rogers, Anupama Sen, and Jonathan Stern have been helpful and supportive. I am particularly grateful to Scott McLachlan, the Institute's librarian, who tracked down elusive books and articles, and to the staff of the British Library, where I did much of the research.

I learned a great deal from presenting at the Sheffield Hallam university history seminar (2014), a Canterbury Christ Church University/UCL seminar (2015), and at the LSE Grantham Research Institute (2017). Presenting at the Historical Materialism conference (2014), the Planetary Natures conference at Binghamton University, USA (2015) and the World Ecology conference at Durham (2016) was also invaluable. I thank, especially, Diana Gildea and Jason Moore for their support at the two latter events. Also, I learned from introducing research material at the Red Green Study Group in London (2015), the Radical Anthropology Group (2015) and the 'social crisis' discussion group (2016), and from conversations with Emma Hughes, Mika Minio-Paluello and James Marriott at Platform London.

Others who have shared their knowledge with me or helped in other ways include Nic Beuret, John Bulaitis, Katya Chertkovskaya, Gareth Dale, Brian Davey, William Dixon, Laura El-Katiri, Don Filtzer, Roger Fouquet, David Goldblatt, Ian Gough, Tim Gould, Amelia Hadfield, Barbara Harriss-White, Nick Hildyard, Jane Hindley, Sohbet Karbuz, David Lamoureux, Larry Lohmann, Brendan Martin, Shonali Pachauri, Walt Patterson, Andrew Pendleton, Andrew Pragacz, Maria Sharmina, Pritam Singh, Lorne Stockman, Dave Temple, Olga Tkach, Steve Thomas and Judith Watson.

Many thanks to Antoine Pesenti, and to the Barry Amiel and Norman Melburn Trust, who made contributions to research costs.

I am also deeply grateful to David Shulman and his colleagues at Pluto Press, and to Hilary Horrocks and Wendy Derose. I have been fortunate to have many friends who have taken an interest in this project: thanks to all of them. More than anyone else, my family – Monika first of all, and Nadine, Yusef, Yohan and Kamil – has supported me and made what I do worthwhile.

Units of measurement

Energy. The main units of measurement of energy in this book are tonnes of oil equivalent (toe) and kilograms of oil equivalent (koe). When discussing global and national energy balances, millions of tonnes of oil equivalent (mtoe) are usually used. Standard conversions between toe and other units of measurement of energy are shown in Table 1.

Table 1   Units of measurement of energy

To:	Terajoule (TJ)	Gigacalorie (Gcal)	Tonnes of oil equivalent (toe)	Gigawatt hour (GWh)
From:	Multiply by:			
TJ	1	238.8	23.88	0.2778
Gcal	0.0041868	1	0.1	0.001163
toe	0.041868	10	1	0.01163
GWh	3.6	860	86	1

Source: IEA website

Oil is often measured in barrels. In this book, as elsewhere, international oil prices are referred to in \$/barrel. 7.33 barrels of oil equivalent = one tonne of oil equivalent.

When coal was the dominant fuel, i.e. at least until the Second World War, tonnes of coal equivalent (tce) was a common unit of measurement. One tce is usually counted as 0.7 toe. In Chapter 1, I have left measurements given in tce in those units.

Electricity delivers energy. The *rate at which energy is generated in the form of electricity* is measured in watts (W), kilowatts (kW), megawatts (MW) or gigawatts (GW). The *amount of energy generated in the form of electricity* is measured as kilowatt-hours (kWh), megawatt-hours (MWh) or gigawatt-hours (GWh). A kilowatt-hour is the amount of energy produced by a 1-kilowatt current in one hour.

The capacities of generating stations and networks are usually measured in MW or GW; the amounts of energy they produce over periods of time in MWh or GWh.

It is common to express large volumes of energy produced as electricity in billion kWh (bn kWh), and I have followed this convention, for example, in Chapter 7. One billion kWh = 1000 GWh = 1 terawatt-hour (TWh).

Vehicle fuel efficiency is usually measured in litres/100 km. In discussion of the regulation of fuel efficiency in the USA, I have used miles per gallon (mpg). 10 mpg = 23.52 litres/100 km; 20 mpg = 11.76 litres/100 km; and 30 mpg = 7.84 litres/100 km.

Acronyms and abbreviations

ABB – ABB (Switzerland), a major engineering company
AC – alternating current
AES – AES Corporation, a major electricity producer
BP – formerly British Petroleum, an international oil company
CCGT – combined cycle gas turbine
CHP – combined heat and power plant
CO_2 – carbon dioxide
EAF – electric arc furnace
EIA – Energy Information Administration (USA)
EPA – Environmental Protection Agency (USA)
FAO – Food and Agriculture Organisation of the United Nations
GDP – gross domestic product
GE – General Electric
GM – General Motors
GW – gigawatts
ICE – internal combustion engine
IEA – International Energy Agency
IMF – International Monetary Fund
IOCs – international oil companies
IPCC – Intergovernmental Panel on Climate Change
IPPs – independent power producers
koe – kilograms of oil equivalent
kW and kWh – kilowatts and kilowatt-hours
LNG – liquefied natural gas
LPG – liquefied petroleum gas
mt – million tonnes
mtoe – million tonnes of oil equivalent
MW and MWh – megawatts and megawatt-hours
NEPA – National Electric Power Authority (Nigeria)
OECD – Organisation for Economic Cooperation and Development
OPEC – Organisation of the Petroleum Exporting Countries
PURPA – Public Utility Regulatory Policies Act (USA, 1978)
PV – photovoltaic cells
SEB – State Electricity Board (India)
SUV – sport utility vehicle
tce – tonnes of coal equivalent

toe – tonnes of oil equivalent
UNDP – United Nations Development Programme
UNEP – United Nations Environment Programme
UNFCCC – United Nations Framework Convention on Climate Change
USSR – Union of Soviet Socialist Republics, or Soviet Union
WMO – World Meteorological Organisation
WRI – World Resources Institute
WTO – World Trade Organisation

Introduction

Each year, fossil fuels (coal, oil and gas) are consumed in ever-greater quantities, despite the danger of global warming, which makes such large-scale consumption unsustainable. The facts of consumption growth are at odds with ever more insistent claims that we are moving to a post-fossil-fuel era. Clearly, the causes of consumption growth are very strong. The purpose of this book is to put them into historical perspective.

The book covers the period since 1950, because it was in the second half of the twentieth century that fossil fuel consumption expanded to levels associated with dangerous global warming. The fossil fuel industries had taken a central place in rich countries' economies long before that, and taken their toll on humans and on the natural environment they live in. Tens of thousands of coal miners were burned, buried alive, gassed, blown up or otherwise killed in the production process. Millions of city dwellers' lives were painfully cut short by coal-related air pollution. But the threats to human society implicit in global warming – including the effects of rising sea levels, ruination of agriculture and the destabilising effects of storms – are on a still greater scale.

The accumulation in the atmosphere of the greenhouse gases that cause warming is foremost among the dangerous impacts of human economic activity on the natural world that have mounted over the past two centuries or so. Other notable impacts include the disruption of the nitrogen cycle and substantial loss of biodiversity. There is a consensus among researchers that these impacts signify that a new geological epoch has been reached: the Anthropocene (i.e. the 'new epoch of humans', from the Greek word *anthropos* (human)). Earth systems scientists, who work on integrated analyses of human-natural relationships, have concluded that since the mid-twentieth century there has been a 'great acceleration' of these impacts, and that fossil fuel use is the most significant cause.[1] The aim of this book is to throw light on this aspect of the problem.

The character of the damage done by excessive fossil fuel use has become clearer in the course of the time period covered by this book. It became apparent to climate scientists in the mid-1980s, and was accepted in international political forums in the early 1990s, that the global warming danger necessitated sharp reductions in the level of fossil fuel consumption. But it has kept rising. It swelled by more than half in the quarter century between 1990 – when the Intergovernmental Panel on Climate Change (IPCC) issued its First Assessment Report, formally urging a strategy to reduce consumption – and 2015.[2] At the Copenhagen climate summit in 2009, the world's most powerful governments failed to agree on such a strategy, and by the Paris summit in 2015 admitted that they *could* not. In a century's time, when the impacts of global warming will be much more ruinous than they are today, people may look back at this failure as collective madness. There may be an analogy with the way that people today view Europe's descent in to the barbaric slaughter of the First World War, a century ago, as collective madness. It *was* madness, but it had definite political, social and economic causes that historians have sought to understand. In this book I will try to do likewise, with the madness that is producing global warming.

Consumption through systems

Fossil fuels are consumed primarily by and through technological, social and economic systems, and these are this book's main focus. Relatively small quantities of coal, oil and gas are consumed directly by individuals and households, such as for heating or cooking, or (in the form of oil products) to fuel their cars. But most fossil fuels are consumed indirectly. They are used in the production of materials – from steel and cement to plastics and fertilisers – for industry and agriculture, which in turn produce goods for consumption; as fuel for industry and for transporting goods; for construction; or for military or other state functions. They are used as fuel to produce electricity or heat, which in turn have multiple uses. Where individuals consume fossil fuels directly in the technological sense – petrol in a family car, for instance – they do so in the context of social and economic systems over which they may have little control – in this case, urban development that sites homes, jobs and shops far from each other, work patterns that require them to make particular journeys, and so on. More broadly, technological systems that consume fossil fuels (e.g. electricity networks) are shaped by the social and economic systems in which they are embedded (e.g., in the late twentieth century, capitalism, or state socialism). Fuels are consumed not by undifferentiated humanity, but by people living in, and divided by, these unequal social and economic systems.

The interpretive approach of this book, focused on these technological, social and economic systems, is at odds with the assumption, shared by many economists, that the function of any economy is essentially to serve consumers' demand. In my view, production and consumption in the global economy have a symbiotic relationship, determined ultimately by relations of wealth and power in the economy. The driving forces for economic expansion lie ultimately in the constant urge of capital to accumulate, that is, for the wealth and power that dominates society to renew and reassert its dominance. (This view, formed in the Marxist tradition, is discussed in Chapter 11.) By making technological, social and economic systems the starting-point, my interpretation also contrasts with some writing about consumption, that concentrates on the cultural and social contexts in which mass consumption has expanded in rich countries over the last century, and on consumers' psychological motivations. Cultural trends, and psychologies, form part of the story, but need to be considered in the context of the social, economic and technological systems.

These systems have since 1950 evolved, especially, through six social and economic processes, to which I shall repeatedly refer: *industrialisation*, and especially the expansion of energy-intensive industries such as steel and cement production; *technological and other changes in the labour process*, both in industry and in the domestic sphere; *electrification*, which was pretty well completed in rich countries during the post-war boom, but continued in most of the world since then; *urbanisation* and *motorisation*; and *household material consumption and the growth of consumerism*.

This book addresses a significant gap in historical literature. There is a rich historiography of the global development of fossil fuel industries, by writers including Vaclav Smil, Janet Ramage, Bruce Podobnik, Matthew Huber, Timothy Mitchell, Andreas Malm and many others.[3] The history of energy consumption is a narrower field, in which I have learned, especially, from the work of David Nye on the USA and Sunila Kale on India.[4] The expansion of fossil fuel consumption in recent decades has been the subject of a mountain of reports by international agencies, economists, energy specialists and NGOs, but much less work, so far, by historians.

Several historians have pointed out the need for a specific treatment of fossil fuel use. We lack 'a history sufficiently precise [...] to distinguish the share of responsibility of different technological choices for the climate crisis', that can identify 'the main institutions that have set us on the road to climate cataclysm', and by which historical processes, Christophe Bonneuil and Jean-Baptiste Fressoz complained. Frank Trentmann concluded his general history of consumption by calling for discussion of its environmental costs. Adam Tooze responded in a review that Trentmann's approach – which views consumption as 'individualistic, creative and cosmopolitan', 'essentially within our control', and subject to politics – did not allow examination of the way consumption is 'crashing against environmental limits'. The history of 'rampant fossil-fuel consumption' had to be addressed; we need 'a history that shows how consumption and production became tied together in an expanding feedback loop of ever greater economic and material scope'.[5] Hopefully this book is a step on that path.

I hope that understanding how fossil fuel consumption has spiralled out of control since the mid-twentieth century will help us all to shape the future transition away from fossil fuels. I cannot say how that will happen, and I am sceptical of those who pretend to be sure. Nor am I neutral about the future, though, and in Chapter 12 I have presented my view of the factors that I think will shape it. A theme running through that chapter is that the decisive actor is society – all of it, collectively – rather than political elites. For this reason I do not offer what are sometimes called 'policy recommendations'. I have tried to balance this general outlook with comments on specific ways in which technological, social and economic systems may change.

Some terms

Fossil fuels are consumed in three main ways. 1. They are consumed directly, e.g. by a steel plant for processing heat, or a household for cooking. 2. They are transformed into other forms of energy (electricity or heat), or into intermediate products (e.g. petrol, kerosene or diesel fuel, all refined from crude oil) that are then consumed by businesses, state bodies or households. Following usual practice, I refer to this as *transformation*. 3. They are used as *raw materials*, e.g. in the petrochemicals industry to produce plastics, other industrial materials or chemical fertilisers for agriculture.

Since 1950, most fossil fuels were produced, transformed and supplied commercially, i.e. by corporations, state bodies and other economic actors. There are also non-fossil-fuel energy sources – hydro power from dams, nuclear power and modern forms of renewable energy (mainly solar and wind power) – that are produced commercially. The *commodification* of energy – i.e. the transformation of coal, oil and gas, their by-products, electricity, and heat into commodities traded in markets – is a theme that runs through the book. I distinguish between *commodified energy* and *energy supplied as a state benefit*; both of these are *commercial energy*.[6]

Hundreds of millions of people, mostly in rural areas outside the rich world, use *non-commercial* forms of energy. The most significant of these are *biofuels*: wood, other biomass and animal dung, collected from the natural environment mostly by the families that consume them directly.

Energy is defined in two main ways. For physicists, energy is the 'ability to do work' – a definition that includes human and animal labour power. That definition is too broad for people, including me, writing about energy in its social context. I describe labour as labour, and energy as *work done by physical or chemical resources, mobilised by people for that purpose*. This follows the Oxford Dictionary definition of energy as 'the means of doing work by utilising matter or radiation'. That 'work' can include anything from running a power station to warming a room. A vital function of energy resources, including fossil fuels, has been to substitute for human labour, whether in industry, agriculture or in households.

A physicist would say that energy can neither be 'produced' or 'consumed', because humans' energy systems simply take energy in one form and change it to another form. Nevertheless I have used these words in the usual way.

Fossil fuels, non-fossil energy sources such as nuclear fuel or solar panels, and manufactured forms of energy such as electricity are all *energy carriers*, i.e.

physical phenomena that carry within themselves the 'ability to do work'. I have used this term when necessary for clarity, but I have also used the term 'energy' to refer collectively to energy carriers.

I use *renewables* to refer to solar, wind and tidal power, but not hydroelectric power, in line with common usage.

In the period after the Second World War, governments, international agencies and research institutions began systematically to compile national and international *energy balances* that counted economies' total energy inputs and outputs.[7]

A final note on terminology: I refer throughout to the *rich world*. It is an unsatisfactory generalisation, covering the countries that were most completely industrialised, and achieved the highest living standards, before the Second World War: the USA and Canada, most of western Europe, Japan and Australia. Often, these countries have become, and stayed, rich at others' expense, exploiting populations and resources through colonialism in the nineteenth and early twentieth centuries, and through economic domination in the period covered by this book. There are some patterns of energy use largely shared by those countries, some of which spread to other countries during the period covered. One way of defining the 'rich world' is to refer to the member countries of the Organisation of Economic Cooperation and Development (OECD). I have sometimes used this marker, but it is not always helpful. (For example the OECD includes Turkey, but not Saudi Arabia, whose per capita gross domestic product (GDP) is two-and-a-half times greater).[8] As for countries outside the rich world, I sometimes refer to them in just that way. In places I have fallen back on the term *developing countries*, while being well aware that it fails to encompass the way many countries' development has been blocked and confounded.

How the book is organised

Part I sets out contexts: historical (Chapter 1), technological (Chapter 2), and social and economic (Chapter 3). Chapter 4 presents a picture of fossil fuel consumption since 1950 as measured statistically. Part II is chronological. Chapter 5 covers the 1950s and 1960s, and Chapter 6, the 1970s. Chapter 7 compares the electrification of some key countries. Chapters 8, 9 and 10 cover, respectively, the 1980s, the 1990s and the 2000s. Part III draws together reflections and conclusions. Chapter 11 summarises the evolution of approaches in social theory to consumption in general, and fossil fuel consumption in particular. Chapter 12 is forward looking, and considers what history might tell us about the transition away from fossil fuels. Chapter 13 presents some conclusions.

Any author of a global history has to decide how much detail to include on any particular country. I have put in the foreground the USA (the twentieth century's largest fossil fuel consumer); China (the largest consumer in the 2010s); India;

Nigeria (Africa's most populous country); South Africa (Africa's most industrialised country); western Europe; and the Soviet Union and successor states. Too little is included about some significant energy consumers (e.g. Japan), and many important countries with huge populations (e.g. in Latin America, the Middle East and South East Asia). I hope this is justified by the clarity added to the overall picture by details about the countries I have focused on.

Part I

Contexts

1
Fossil fuels before 1950

The history of human consumption of fossil fuels can be divided into four time periods:

1. Human history before the European Industrial Revolution, when, apart from some local, temporary episodes, fossil fuels played no significant part in economic activity.
2. From the start of the Industrial Revolution in the mid-eighteenth century up to about 1870, when coal mining, coal-fired steam power and coke-fuelled iron making took centre stage.
3. From 1870 to the mid-twentieth century, when the second Industrial Revolution, fuelled by coal and to a lesser extent oil and gas, produced electricity networks, automated manufacturing, the internal combustion engine and petrochemicals. Such fossil-fuel-dependent systems became central to rich countries' economies.
4. From the mid-twentieth century to the present, when fossil fuel consumption expanded to many times its previous levels, fossil-fuel-dependent systems expanded outside the rich world, and oil surpassed coal as the most widely used fuel.

This chapter covers time periods 1–3. The fourth period is the subject of the book as a whole. Figure 1 shows how the use of fossil fuels has grown dramatically in periods 3 and 4.

In periods 2, 3 and 4, or since the mid-eighteenth century, a new relationship between human society and its natural surroundings has taken shape. The impacts of human activity on the earth and its natural systems have begun to operate on the same, or greater, scale as those systems themselves. These impacts include: destruction by agriculture and industry of biodiversity (the extinction of species at an unprecedented rate); disruption of the nitrogen cycle (the circulation of nitrogen through air, soil and water); and the acidification of oceans. But the most significant impact is the change to the atmosphere's chemical composition through the release of greenhouse gases – and the main cause of this is the burning of fossil fuels, which emits carbon dioxide (CO_2). (See pp. 56–8.)

A consensus has formed between researchers, and many other people, that we therefore now live in a new geological epoch, the Anthropocene – as distinct from the Holocene that began at the end of the last Ice Age.[1] The exact dating

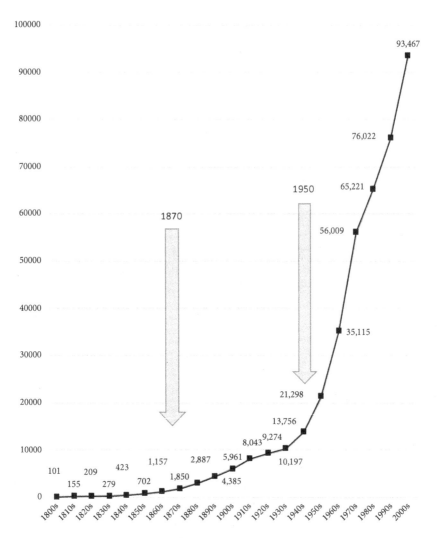

Figure 1 Fossil fuel production, 1800–2009
Million tonnes of oil equivalent per decade

of the Anthropocene epoch, and other aspects of the concept, are subjects of controversy. But the natural scientists are clear, collectively, that there was a sharp upturn in the whole range of human impacts on natural systems from the mid-twentieth century – period 4 referred to above.

From the beginnings to the Industrial Revolution (before 1870)

For thousands of years, until the eighteenth century, human and animal labour power were the main sources of energy for economic activity. Water wheels and windmills were used as prime movers (converters of source energy into mechanical energy), but much more energy was expended by domesticated animals, such as horses, oxen or donkeys, and humans themselves. Fields were ploughed, barges and carts pulled, treadmills worked and bellows operated by horses, slaves, serfs or free people.

During those thousands of years, people burned coal to produce heat and light, but surface outcroppings were rare, and coal churned out smoke. Plentiful wood was the main fuel. In some places, people used oil from the ground as a medicine or lubricant, but vegetable oils and animal fats were preferred. The first use of coal on an industrial scale was in China: in the eleventh century, shafts were sunk to mine it, and it was used extensively for metallurgy. The reasons why the Industrial Revolution did not happen in China have long been debated among historians; certainly the absence of a coal-fired prime mover were among the constraints, and the distances between coal deposits and urban settlements. From the eleventh century onwards, surface deposits of coal were mined in Scotland, England, Belgium and France, and used for forging iron, lime manufacture and evaporating seawater to prepare salt. But wood was the dominant fuel. Historians of London, for example, have shown that, even in the sixteenth century, when wood supplies were squeezed and prices rose, the additional cost of transporting coal to the city made it uncompetitive. The first European country in which a fossil fuel became dominant was the Netherlands, where peat largely replaced wood in the seventeenth century.[2]

It took the Industrial Revolution – the triumph of mechanised factories over workshops, of iron over other materials, and the rise of steam power – to give coal a central position in the economy. Coal's main function, in economic terms, was to substitute for animate – human and animal – labour power, vastly increasing industrial productivity. This transformation, which began in Britain in 1750–1830, was not only technological, but social. Capitalist wage-labour, which had roots going back centuries in English agriculture, took on a central role. Money made from the transatlantic trade, and slavery in America, helped Britain to finance the Industrial Revolution; that revolution, in turn, reinforced British supremacy over world trade and colonialism.[3]

Technologically, the Industrial Revolution started not with coal and steam but with the mechanisation of cotton manufacture. In the second half of the eighteenth century, mechanical spinning jennies, water frames and power looms were introduced and factory-based cotton manufacture soared, leaving behind the workshop-centred wool industry. The two crucial techniques that boosted coal demand – coke for iron making, and the steam engine – had both been used since the beginning of the eighteenth century, but were widely diffused only towards its end. Coke, made from coal – which in Britain was abundant and cheap – was burned in blast furnaces instead of charcoal. This cut the cost of making iron, which served as raw material for machines, farm tools, water and gas pipes, and weapons of war. In the 1780s Britain made less iron than France; by the mid-nineteenth century it was making more than the rest of the world put together. The steam engine was the first machine that converted fossil energy resources into mechanical work, and not just heat. Concentrated volumes of mechanical power, previously available only from strong natural water flows, could be unleashed almost anywhere. The engine invented by Thomas Newcomen in 1705 was already used widely to pump water from mines by the mid- eighteenth century; it was James Watt's crucial improvement, the addition of a separate condenser, first applied commercially in 1776, that brought steam engines into general use. They became more fuel efficient and more adaptable, for use in factories, trains and ships.[4]

Steam and iron drove coal's dizzying expansion in Britain in the nineteenth century – but coal had begun to compete with wood long before that. Already in 1700, coal had overtaken wood as a source of thermal energy in Britain. The reasons that coal won out have been the subject of controversy among historians. The natural constraint on wood production, that Britain had only a fixed amount of land on which to grow it, was the focus of Edward Wrigley's analysis. Without coal and the shift from a wood-fuelled 'organic economy' to a 'mineral-based energy economy', he argued, economic growth would have faltered. Other historians were unconvinced: there *were* shortages of wood, but these were local (especially in densely populated areas) and transitory. Transport also made a difference: wood had to be collected from multiple locations, and was bulkier to move. Coal, despite being much dirtier to burn, got a foothold in such industries as pottery, brick- and glass-making, as well as iron-making.[5] But it was coal's sheer abundance and its ability to substitute for human labour that were decisive, argued Robert Allen. In eighteenth century Britain, the blast furnace, steam engine, spinning jenny and water frame increased the use of coal and capital relative to labour. These technologies were adopted and diffused in Britain more rapidly than elsewhere, because 'wages were remarkably high, and energy was remarkably cheap'.[6]

In the nineteenth century, steam engines became the leading consumers of coal. But they did not overtake wind and water power overnight. Early steam engines were very expensive and inefficient by any standards. The earliest ones had thermal conversion efficiencies of 1 per cent (the output was 1 per cent of the energy content of the fuel input) and it took a century to boost this to around 20 per cent.[7] Even with Watt's improvements, the engines were relatively inefficient, and wind and water remained dominant in industry. Steam had obvious advantages, though: cotton mills no longer had to be located near flowing water. Employers used the new technologies to reshape their social relations with workers. Factories could now be sited where employers could best force workers into them and best control them while at work. During a series of labour revolts culminating in the 1842 general strike, workers acted against the machines, as the Luddites had a quarter of a century before, and disabled engine houses and pitheads. 'This was collective bargaining by rioting against the fossil economy,' Andreas Malm argued, coining Eric Hobsbawm's phrase.[8]

In 1800, Britain's coal consumption was 11–15 million tonnes (mt)/year; by 1845, 40–45 mt; by 1870, it was crossing the 100 mt mark. The spillover of steam engines into railways gave a further impetus to industrial development. Coal for steam engines, and wrought iron for rails and wheels, underpinned the British railway boom of the 1830s and 1840s. That boom in turn made coal more easily transportable, boosting its competitiveness against wood. Steam was also introduced into ships, but replaced sail only slowly – the US merchant fleet, for example, was 15 per cent steam powered in 1850 and 33 per cent by 1880.[9]

The British Industrial Revolution also led to unprecedented urban development. There had been many cities, including some very large ones, in world history. But industrial cities – populated by wageworkers and their families, full of factories, with streets underfoot and air above full of smoke and soot – had never existed on this scale. By 1860, 50 per cent of England and Wales was urbanised, compared to 25 per cent of Italy, Belgium and the Netherlands and 18 per cent of France. In nineteenth century Britain, the lighting of streets and factories made it possible to lengthen the working day; it went together with clean water and sewage systems designed to minimise the effect of regular and dreadful epidemics. In the USA, too, in the nineteenth century people began to receive water, gas and some steam heat from sources outside the home, well before electrification. This made it easier for male workers to be separated from the daily routine of work at home and to go to the factories. In the eighteenth century, municipal lighting had often used whale oil or vegetable oil; in the nineteenth century, increasingly, coal gas (methane recovered from coal).[10] Up to the 1870s, though, consumption of fossil fuels in people's homes was rare and statistically insignificant.

Coal and steam 'did not make the industrial revolution, but they permitted its extraordinary development and diffusion', historian David Landes pointed out.[11] From about 1830, the coal- and steam-based industrial system spread to France, Belgium, and to the states that would be unified in Germany in 1871. There followed a new round of colonisation led by Britain. As Bruce Podobnik wrote:

> Coal-powered ships and railroads allowed Britain and its Continental rivals to seize control over territories in Asia, Africa and the Middle East that had long resisted conquest. Coal-driven transport systems then allowed for a radical increase in the volume of goods moved from the periphery into the core of the world-economy.[12]

The USA expanded too, using slave labour and expropriating native Americans. It was unified by the Civil War of 1861–65, which new weapons technology made the world's most destructive conflict up to that time. Most of the energy for industrial production in these economies was by then provided from coal, although hydro power also made a contribution. The European empires encouraged often one-sided industrialisation in territories they controlled. This, and autonomous development for example in Japan, made coal a truly worldwide industry by the end of the nineteenth century, although even then coal consumption was concentrated overwhelmingly in the rich countries.[13]

The second Industrial Revolution (1870–1913)

Between 1870 and 1913, the second Industrial Revolution produced innovations that underpinned new fossil-fuel-based technological systems. Two of these – electricity networks and the internal combustion engine (ICE) used in cars, trucks, ships and later, planes – still today account *directly* for more than half of global fossil fuel use.[14] Coal consolidated its role in the main capitalist countries. The ICE became the prime source of demand for oil, which began to be consumed in significant quantities in the early twentieth century. Large-scale oil production gave rise, in turn, to the petrochemical industry, which would stimulate the use of chemical fertilisers for agriculture. Further indirect consequences of the second Industrial Revolution were manifested throughout the twentieth century, in the types of cities in which much of the rich countries' populations would live, in the industries in which they worked, and in the consumer goods they would buy.

World coal output during the second Industrial Revolution dwarfed that of the first Industrial Revolution; it rose sixfold between 1870 and 1913. (See Figure 2.) By the turn of the century there were modern mines operating on all continents, but the lion's share of production and consumption was in Britain, France, Germany and the USA. In the last quarter of the nineteenth century, coal consumption in the USA surpassed the total for western Europe and the USA became the largest consumer nation. Oil and gas production, which was negligible in the nineteenth century, was estimated jointly at 37.4 million tonnes of coal equivalent (tce) in 1900 and 96.6 million tce in 1913.[15]

Electricity, the first significant technology of the second Industrial Revolution, had first been generated for industrial applications by hand-driven dynamos in the 1830s, a few years after Michael Faraday's 1831 discovery of the relationship between electric current, magnetism and force. In the 1840s, larger generators were turned by waterwheels or steam engines. The first big source of demand, in the 1870s, was electric lighting for streets, department stores, theatres and the homes of the rich – but it had to compete with well-established gas lighting systems. The invention of electric light bulbs in 1879 by Thomas Edison in the

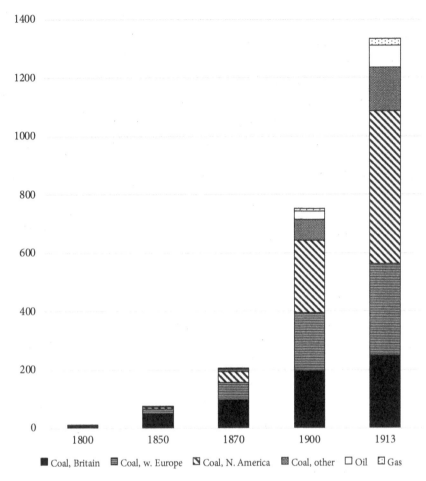

Figure 2 Fossil fuel production in the first and second industrial revolutions
Millions of tonnes of coal equivalent

USA and Joseph Swan in Britain provided the impulse for the first networks, which opened in London and New York in 1882. The generators for these systems were powered by large steam engines. The much lighter and smaller steam turbine, patented by Charles Parsons in 1884, was a crucial technological breakthrough:[16] successors of Parsons's engine have generated most of the world's electricity since then. Edison's first generators produced 90 kW; within thirty years turbine-powered generators were often 100 times more powerful. In Britain the average size of generators installed rose from 500 kW in 1895 to 5 MW in 1913; in Chicago, a 35 MW generator was installed in 1910.[17]

The first electricity networks were local, each plant usually serving a factory or urban area. Large cities had patchworks of isolated systems: by 1900, London had ten different frequencies, 32 voltage levels and 70 different ways of charging for electricity. Networks were unified with the help of transformers and alternating current (AC) transmission systems that could carry electricity over long distances. On the eve of the First World War in 1914, globally there were 55 high-voltage transmission lines, between 30 and 400 km long, mostly in the USA. National electricity networks underpinned the development of the telegraph and telephone, revolutionising communications.[18]

Fossil fuels dominated urban electricity systems from the start. In the countryside, the picture was mixed. Wind turbines began to be used in Denmark from the 1890s and in other European countries soon afterwards. In the USA, hundreds of thousands of farms installed wind turbines in the 1920s–1930s, before they had access to network electricity.[19] In the interwar period, and again during the post-war boom, the growth of centralised networks would push wind power to the margins.

As electricity networks proliferated, the question arose of who should finance, build and, ultimately, control them. On one side were privately owned electricity corporations, who viewed electricity as a commodity to be marketed. On the other were liberals and socialists, who saw electricity, along with gas, water and sewage systems, as a service that should be provided to the urban population by local government. In the middle were national governments for whom electricity provision became a strategic infrastructure problem.

In the USA, the initiative lay with the corporations and their banks. Edison, named after its founder, installed city networks and sold electrical equipment. In 1892 Edison, financed primarily by J. P. Morgan, merged with Houston-Thomson to form General Electric (GE). A duopoly of GE and Westinghouse, first mover in AC transmission, soon dominated the market. By 1893, 15 smaller electrical equipment manufacturers had been swallowed by these two giants. In the late 1890s a third business empire was taking shape, headed by Samuel Insull. Publicly owned systems were also installed, but they had no ready source of capital; the private companies, with access to bank and stock finance, leapt ahead, forming regional networks that went across state and city boundaries. Insull excelled at this type of expansion, using holding companies to hoover up and unify smaller networks. He pioneered the concept of 'natural monopoly', arguing that private corporations like his should be permitted to supply geographical areas as a state-regulated monopoly. Feeble regulation and institutionalised corporate influence were born.[20]

In Europe, the idea that electricity was a public service, and that the state should ensure its widespread provision, was stronger. During the British Industrial Revolution, as factory workers and their families had been packed into crowded, smoke-filled slums, public health and municipal services had become a burning political issue. By the late nineteenth century most British cities were providing water and sewage systems, gas lighting, refuse collection, and public health services, paid for by rates. When electricity appeared in the

1880s, municipalities treated it as another service for urban populations; this converged with national governments' anxiety to provide the necessary infrastructure for this revolutionary technology. British municipalities were required by law to lay cables.[21] In Germany, these services were at first provided by private business, but in the wake of bankruptcies, or to meet public needs, municipalities often took over. In France, the market was dominated by privately owned regional networks, but municipal provision was established by law in 1903, and later a system of concessions developed.[22]

Industry was the prime user of electricity in the cities that had it. Electricity became the prime source of mechanical drive in factories. Motors could be fitted to tools, and the tools moved to the job. Larger machines that had previously had to be stationed near water wheels could be arranged differently. Automation followed. In US factories between 1899 and 1929, total installed mechanical power quadrupled, and electricity's share rose from 5 per cent to 82 per cent. Thanks to the shift from raw energy forms such as coal and water power to electricity, output per worker-hour and per unit of capital input rose steadily. Industry was also the main electricity consumer in Europe: in Italy, at the turn of the century, more than four-fifths of generators belonged to factories.[23]

Electricity was rare in households before the First World War. Even by 1910, only one-tenth of US homes were wired. Insull's aggressive profiteering played a role in developing residential consumption. His Chicago company focused on raising the number of households attached to the grid: it introduced electricity meters and charged for the amount consumed rather than per light bulb, which meant lower bills. In a famous advertising campaign it gave away 10,000 electric irons to new customers. Irons, lights and radios were the three most widely used electrical appliances before the First World War.[24]

The second Industrial Revolution stimulated fossil fuel consumption in many other ways apart from electricity generation. Steelmaking – for railways, ship hulls, bridges, buildings, and all types of tools and machinery – drove up coal use. Transportation was transformed: railways proliferated, steam trains got bigger and faster, and sailing ships were superseded by steam ships. The ICE was, after electricity, the most significant new technology. The first light, high-speed petrol engine was patented by Gottlieb Daimler in 1885, and in the same year, Karl Benz built the word's first car. Daimler's engine, in which the fuel was ignited by a spark, was followed by Rudolf Diesel's engine, in which the fuel ignited spontaneously; this was inherently more efficient and used heavier liquid fuels. Shortly before the First World War, ICEs, half the weight or less of the best steam engines producing the same motive force, began to be used in road vehicles, ships and railway engines. Substituting for labour was a big consideration: running a ship on oil meant that you could dispense with coal stokers, who made up half the crew. Once the Borneo and Texas oil fields started producing, in 1898 and 1901 respectively, a steady supply of the new fuel was available and car manufacture began.[25]

From the First World War to the Second World War (1914–40)

Germany, France, Belgium and the USA had used the advantage afforded by coal-based industry and transport to build empires that could compete with Britain's. Together with Britain, they brought money, manufactured goods and great human suffering to their colonies in Africa, Asia, the Middle East and Latin America. Rivalries between the capitalist powers engulfed Europe in war in 1914–18. The old, largely pre-industrial empires – Russian, Ottoman and Austro-Hungarian – collapsed as a result. In the imperial centres, industry had given rise to a working class, whose organisations were, by the First World War, challenging for political power. The Russian revolutions of 1917, which brought to power a party claiming to speak for working people, sent shivers down the spines of old authorities everywhere.

After the war, a decade of economic recovery was followed by a deep recession, from 1929. Thereafter the USA rose to become the dominant economic power and, after the Second World War of 1939–45, the dominant political one. Global fossil fuel consumption doubled between 1913 and 1950. It reflected the geopolitical shifts: the USA's consumption rose far more rapidly than the European powers'. In 1925, North America (i.e. the USA and a small amount for Canada) accounted for just over half of global fossil fuel consumption; in 1938 it dipped to just under two-fifths; by 1950 it was again around half. In Europe, consumption fell after the First World War, rose during the Depression and fell again during the Second World War. Fuel consumption grew fastest in the USSR – nearly twelve times over in the quarter century 1925–50, largely as a result of forced industrialisation. (See Table 2.)

On the eve of the First World War, in 1913, oil and oil products accounted for less than 6 per cent of the world's commercial energy balance, and gas 1.5 per cent. The war forced forward the development of oil-based transport. Generals had planned to fight it with horses, but by 1915 cars, trucks, tanks, aeroplanes and submarines had become decisive. Between the wars, oil consumption continued to rise rapidly. Refining technology developed: thermal cracking was widely diffused from the 1920s and catalytic cracking from the 1940s.[26] The most important driver of oil demand was motor cars, most of which were in the USA. Diesel began gradually to displace coal-fired steam, both in shipping and on railroads. But in sheer volume terms, coal remained dominant until after the Second World War. As shares of the global commercial energy balance, in 1925 coal was 83 per cent, oil and gas 16 per cent; in 1950, coal was 61 per cent, oil and gas 37 per cent. Coal dominated power generation, fuelled industry, and had an almost complete grip on space heating provision in rich countries.[27]

Table 2 Commercial energy consumption by major world regions and by source, 1913–68

Millions of tonnes of coal equivalent

	1913 est.	1925	1938	1950	1960	1968
World	1334.6	1484.5	1790.1	2610.9	4196.1	6306
By region						
North America	594.1	748.9	706.9	1276.3	1659.5	2359
Latin America	6.7	24.7	38.7	66.2	153.5	245
Western Europe	590.1	517	619.2	583.9	849.5	1242
Eastern Europe	40	55.1	67.7	160.8	293.9	408
USSR	42.4	25.3	176.3	303.3	640.6	1025
Japan	20	30.5	62.4	45.8	111	280
China		23.7	27.3	43.1	235.3	332
Rest of Asia	22.3	29.8	50	60	136.7	242
Oceania	11.7	15.6	18.1	29.3	45.8	72
Africa	7.4	13.9	23.4	42	70.2	102
By source						
Coal etc	1235.7	1230	1291.8	1593.2	1998.5	2315
Oil and products	76.1	196.7	375.8	722.2	1499	2702
Natural gas	20.5	47.9	99.7	252.1	612.6	1157
Hydro electricity	1.7	9.8	22.8	43.4	86	132

Source: Darmstadter, *Energy in the World Economy*, p. 10 and p. 13 (for 1925-68); author's estimates, based on Etemad & Luciani, *World Energy Production* (for 1913)

Notes: The row 'coal' includes very small quantities of peat. For 1913, all of Europe is counted together and all of Asia is counted together. For 1950 and 1960, the row China includes small quantities of fuel for N Korea and Mongolia.

Between the wars, oil imports to the rich world from elsewhere began to dominate energy trade. Net oil exports from underdeveloped countries (as measured by the UN, including Latin America, Asia and the Middle East) to rich countries rose from 27.1 million tce in 1929 to 158.4 million tce in 1950. The short-haul coal trade remained significant, accounting for most of the 300 million tce of intra-regional trade in 1949, while coal exports from rich countries – which had been significant in the imperialist expansions of the late nineteenth century – fell, from 23 million tce in 1929 to just over half of that in 1950.[28]

During the interwar period, international agencies and researchers began systematically to gather statistical records of global fuel consumption. World consumption of commercial energy, broken down by region and fuel, is shown in Table 2. This table, based on the findings of a US-based research project in the 1960s, does not include non-commercial fuels – mostly fuel wood and other biomass used in the countryside in poor countries. United Nations researchers estimated that in 1949, these non-commercial sources amounted to 592.3 million tce, just short of one-fifth of the total, while commercial energy supplies totalled 2,512.3 million tce.[29] Estimates of fuel consumption per person in 1925–68 are shown in Figure 10 on p. 208. The yawning gap between the USA and other rich countries, and the gap between them and the rest of the world, is evident. In the quarter century 1925–50, per-head consumption in China grew by nearly half, but still remained lower than one-hundredth of the US level. (On consumption-per-head statistics, see pp. 50–2.)

The two largest sectoral drivers of fossil fuel consumption growth between the wars were electricity networks and road transport. In the rich countries, electricity consumption increased four times over between 1920 and 1939; industry remained the largest consumer. Electricity continued to replace steam and water as the main source of power; electric lights replaced other types. It was essential for assembly lines and other automated systems, associated with the Taylorist and Fordist systems of work discipline. Line assembly, first used in Ford's Detroit factory, where it was installed in 1910–14, and then to make small arms during the First World War, spread rapidly thereafter to the manufacture of e.g. sewing machines, bicycles and typewriters.

In people's homes, new appliances – most significantly the radio, and devices that eased the burden of domestic labour such as refrigerators and washing machines – drove demand growth. The USA was far ahead of other rich countries: by 1929, more than 1 million homes had electric refrigerators, and by 1941, 16 million. US manufacturers also led the way in aggressive marketing techniques. Before and after the First World War, air conditioning manufacturers battled furiously with engineers, and New York state regulators, who argued that schools would serve their pupils' health better with fresh air from open windows than conditioned air. Elsewhere, big leaps in residential electricity use did not come until after the Second World War. In Europe in the 1940s, most homes' entry circuits carried, at most, 10 amps (the amount of current used by one twenty-first century dishwasher or toaster); the most electricity-intensive appliance was usually the electric iron.[30]

The opposing views of electricity's social function, as a commodity or as a service, influenced the way that networks expanded. In Germany, the Netherlands and Scandinavian countries, where the public service ethos was strong, electrification of every home was seen as a political goal, and would be 90 per cent achieved by 1930. Even in the countryside – which is inherently more expensive to electrify – roughly two-thirds of homes in these countries were wired by 1920. In the USA, where the pace of electrification depended on private companies' willingness to invest, the rural electrification rate stayed under 10 per cent in the 1920s.[31]

The economic depression, and state-led investment policies adopted to address it, brought challenges to the commodity approach. The 1929 financial crash had left Samuel Insull, the electricity distribution pioneer, unable to roll over the huge debt portfolio on which his empire depended. In 1932 it collapsed. Insull was charged with fraud and fled the USA in disgrace.[32] The network of holding companies he had used to expand his business would be effectively destroyed by the 1935 Public Utility Holding Act. Electricity strategy became a big issue in the 1932 election, held soon after Insull's bankruptcy; Franklin Roosevelt, who advocated expanding publicly owned power systems as part of his New Deal to revive the economy, won the presidency. Rural electrification was undertaken through the Rural Electrification Agency, a government agency that financed rural cooperative electricity projects and state infrastructure companies including the Tennessee Valley Authority. The cooperatives played a significant part in rural electricity supply, and by 1948 managed electricity supply to 1.9 million farms.[33]

In the Soviet Union, industry was transformed by an ambitious state-financed electrification project, the largest yet outside the rich world. While for the western European municipal socialists, electrification was a means for local government and/or cooperatives to constrain the power of private enterprise, for the Soviet leadership, it was the key to modernisation by the socialist state on a national level. The Bolsheviks, having consolidated their power with victory in the civil war, prioritised electrification during both the New Economic Policy period (1921–28) and the forced industrialisation drive that followed. By 1940, electricity output was more than 20 times the 1913 level. Such state-financed expansion would come to be seen as a model by developing-country governments. (See pp. 107–21.)

The motor car – an economic, social and cultural phenomenon no less significant than electricity – first rose to prominence in the USA in the interwar period. The shift to large-scale road transport, widespread car ownership and car-based urban development profoundly changed US society. The car-centred transport system became a template for the rest of the world and shaped fossil fuel consumption patterns internationally. Before the First World War, the USA manufactured tens of thousands of cars per year; in the 1920s, millions. By 1929 there were an estimated 32 million cars in the world, about three quarters of them in the USA. For several years thereafter, car sales plummeted, due to the Great Depression, but car ownership levels did not. The industry concentrated as it expanded: there were 88 carmakers in 1921, 43 in 1926 and ten in 1935. By then, the big three manufacturers – Ford, General Motors (GM) and Chrysler – accounted for more than 90 per cent of sales and comprised one of the world's most powerful corporate political lobbies.[34]

Historians trace the motor industry's lobbying power back at least as far as the 1916 and 1921 federal road acts, which assigned government funds for road construction over the vigorous protests of railway companies that had dominated long-distance transport until then. After a wave of closures in the early 1920s, the railway owners fought a rearguard action, absorbing independent bus companies and growing their freight capacity eightfold. General Motors responded, going into the intercity bus business in 1928 and supporting the formation of the Greyhound Corporation. Inter-urban trolley lines, which in 1917 had more than 70,000 km of track, were bought up by a GM affiliate and substituted by GM-built diesel buses, sold with contractual prohibitions on investment in electric trolleys. The 1930s recession, and renewed investment in rural road building under a further act of 1928, sealed the railways' fate: between 1939 and 1972 they lost half of their passenger traffic and nearly three-quarters of their freight revenue, as the road system expanded remorselessly.[35]

The battle over transport within cities was fought still more desperately. Until the 1920s, intra-city transport had been dominated by public transit systems, usually electric. The motor industry sought to replace these with a system based on motor buses and private cars. A holding company, set up by GM together with Firestone Tire, Standard Oil and Mack Truck, between 1936 and 1945 bought up and shut down 100 streetcar systems in 45 US cities. GM signed contracts with local transit companies preventing the purchase of equipment using any fuel except petrol. By the 1950s public transit was in terminal decline, and by 1960 had almost disappeared. Historians have argued about cause and effect in its demise. For some, the investment strategies of oil companies and car manufacturers simply eroded the public transit system to the point where it became economically unviable. Others emphasise that the final blow was delivered by a conspiracy. This latter view gained credence from the 1949 decision of a federal jury to convict GM and its allies of conspiring to dismantle trolley lines. (They were fined a risible $5,000.)[36]

The evolution of car-based cities in the USA depended on road construction. The carmakers collectively became aware in the late 1930s that the level of car ownership in cities was substantially lower than in the countryside, and that 90 per cent of trips were on 11 per cent of out-of-town roadway. They set out to change this. An industry lobbyist wrote in 1939 that drivers did not want to bypass cities, and that for cars to be used fully, 'cities must be remade'; city dwellers who 'refuse to own cars', or use them very little, were 'the greatest untapped field of potential customers'; and road builders should 'dream of gashing our way ruthlessly through built-up sections of overcrowded cities'.[37] This is a fair description of the direction then taken: a campaign to build urban freeways ensued. Up to half of cities' land area became dedicated to roads, driveways, parking lots and service stations. The car became 'the central American consumer good and the engine of the American economy', the historian David Nye pointed out.[38]

Car-centred transport put into millions of people's hands a machine that increased their sense of personal freedom, and simultaneously hiked their level of direct consumption of fossil fuels. In the early 1920s Ford, followed by the other manufacturers, aimed cars at the mass market: the price of a new one fell from $800 in 1909 to $300 in 1924. In that year Alfred Sloan, then chief executive of GM, originated the annual style change, which tweaked cars' cosmetic features without substantial alterations. It was perhaps the largest-scale effort to that date to sell a product on appearance and feel-good factors. Historians of marketing attribute the concept of 'planned obsolescence' to Sloan: it spread from motoring to other industries, playing a key part in corporate strategies to stimulate consumer demand during the depression.[39]

The Second World War and afterwards (1941–1950)

Trends that underpinned fossil fuel consumption growth in the post-war boom, including the rapid rise of oil, were given a powerful boost by the war itself. From late 1941, all the major combatants poured all the fuel they could into production of armaments, war supplies and other goods needed for the home fronts. Germany was constrained by shortages of both coal and oil; so was the Soviet Union, where the industries moved to the Urals in 1941 relied heavily on wood and peat to make up for it. The Allies thought they could out-produce the Axis, and they were right, the historian Alan Milward concluded; their output of weapons and equipment in 1944 was three times greater. The crucial contribution was made by the USA, where war production rose from 2 per cent of the gross national product in 1939 to 40 per cent in 1944. A laudatory volume about GE's contribution to the war effort said that US industry had 'smash[ed] bottlenecks' with the help of GE motors and transformers; electricity supply capacity in industry, per worker, more than trebled during the war. One researcher estimated that the war's overall share of US energy consumption was 40 per cent at its height; another, that the US defence department's share rose from 1 per cent in 1940 to 29 per cent in 1945.[40]

The war not only burned an ocean of fuel, but also set the scene for the US and other governments to boost the development of fossil-fuel-based technologies and infrastructure, in particular aviation and shipping. Technologies including the jet engine and the radar, first diffused during the war, enhanced aviation's post-war development. In 1944 the Chicago Convention, an international treaty, ruled that aviation fuel would not be taxed – a principle then written into 4,000 other bilateral agreements. This huge subsidy to aviation persists to this day. In the immediate post-war years, the US government invested heavily in military aviation; oil-based infrastructure including roads and pipelines to support a network of domestic military bases; an archipelago of overseas bases that by 1990 had a population of half a million; and in shipping and transport. The US merchant fleet had more than tripled in size in the decade to 1947, while those of European nations shrank.[41] There was US state support, too, for fossil-fuel-based fertiliser production: in 1947, the US government turned over to it one of its largest munitions plants, and leftover stocks of ammonium nitrate, a key ingredient for both explosives and fertilisers.[42]

By 1950, social, economic and technological systems to which fossil fuel consumption was central were dominant in the USA and widespread in other rich countries. From 1950 they would expand to dominate on a world scale.

2
Energy technologies

Large-scale fossil fuel use began in order to supply the technologies of the first and second Industrial Revolutions. Since 1950, most consumption was accounted for by the technologies of the second Industrial Revolution, and their spin-offs and successors. To these were added some twentieth century technologies associated with the oil and gas industry, e.g. petrochemicals and plastics manufacture, and the gas turbine. The third Industrial Revolution of the late twentieth century (computerisation and information technology), which had been expected to help decrease consumption, produced in the 2000s big new sources of electricity demand. In the last 30 years, a new question has arisen: how can technologies be deployed to speed the transition *away* from fossil fuels necessitated by global warming? In this chapter I first introduce concepts used to show how energy flows through technological systems; second, describe the most important such systems that have since 1950 produced, transformed and consumed fossil fuels; third, look at the use of technology for reducing fuel consumption; and, fourth, comment on non-fossil-fuel energy technologies and the third Industrial Revolution.

Energy flows through systems

Human societies' energy systems take energy in one form and transform it into other forms. These systems evolved from burning wood for heat, to turning the motive power in wind and water into rotary power, to transforming the more concentrated stores of energy in coal into rotary power with steam engines. By the twentieth century, the systems had become highly complex: coal or oil would be mined in one place, refined in another, and bought and sold several times before being burned in a third; much of these fuels was used to produce electricity or heat; either raw fuels, or electricity or heat, might go into industry, or urban heating and lighting systems, or people's homes. The energy might then undergo further transformation by manufacturing systems, domestic appliances or road vehicles.

Energy systems are technologies – groups of tools and machines used in the labour process by which humans take from nature the means of subsistence and the basis for their society and culture. In this way, these technologies are embedded in social and economic systems. The interaction with the natural world is two-way: fossil fuels are taken from it, while greenhouse gases and other forms of pollution are put back into it.

Methods of energy accounting to track energy flows through technological systems have been around since the nineteenth century. They received a boost in the 1970s, when rising oil prices pushed rich-country elites to find ways of reducing fuel costs and dependence on imports, and funds poured in to the study of energy systems. Researchers adopted conventions set out in Figure 3. These defined four distinct types of energy:

- *Primary energy*. Raw material in the first form in which people obtain it: oil from a well, coal at the pithead, and so on. These fuels may be used immediately, or after some limited processing (as coal and natural gas often are), or may be converted (transformed) into other energy carriers (coal or gas into electricity and/or heat, or oil in a refinery to oil products).
- *Final energy*. Energy in the forms in which it is usually sold to users (electricity, petrol or district heating) for further conversion into useful energy.
- *Useful energy*. Energy in the form that a user wants it: heat for heating or cooking, light for lighting, the motive power of a car, or mechanical energy for manufacturing.
- *Energy services*. The purposes for which people consume energy: keeping warm, being able to see after dark, communication, getting from place to place, or manufacturing things.

These definitions reflect the economic mechanisms of the commodified energy system, and its technological structure.[1] The distinctions between primary, final and useful energy would be meaningless to hundreds of millions of users of biofuels that are mostly not commodified. (See pp. 40–1.)

Energy conversions inevitably involve losses. For example, the laws of physics dictate that the boilers and steam turbines used to produce most electricity lose more than half the energy content as they do so. (See p. 28, *Production and transformation technologies*.) More energy is lost in transport from place to place. In 2000, more than 9,500 million toe (tonnes of oil equivalent) of primary energy went into energy systems globally; after conversion to final energy and then useful energy, around 3,500 million toe of useful energy remained, giving a global conversion efficiency of 37 per cent, according to research by the United Nations Development Programme (UNDP).[2] There are further losses in the conversion of useful energy to energy services.

The term *energy conservation* refers to (a) reducing conversion losses at any point in the flow, such as in electricity transport or industrial processes, (b) cutting final energy requirements by insulating buildings, reducing waste or recycling products, or (c) managing with fewer energy services. The term *energy efficiency* is often used synonymously; in this book it is used to mean increasing

Energy system							
Fossil fuels					**Other sources**		
Extraction and treatment	Oil well	Gas well	Gas well	Coal mine		Uranium mine	
Primary energy	Oil	Natural gas	Natural gas	Coal, lignite	Sunlight	Uranium	
Conversion technologies	Refinery		Heat production	Power plant, cogeneration	Photovoltaic cell	Power plant	
Distribution technologies	Rail, pipeline	Gas grid	District heat network	Electricity grid	Electricity grid	Electricity grid	
Final energy	Petrol	Natural gas	District heat (hot water, steam)	Electricity	Electricity	Electricity	
Conversion of final energy	Cars	Oven, boiler	Local distribution	Electric arc furnace	Light bulb, TV set	Refrigerator	
Useful energy	Acceleration, overcoming air resistance	Heat from radiators	Cooling, heating	Melting heat	Light emission	Cooling	
Technology producing the energy service	Type of car	Building, house, factory	Building, cold storage	Furnace	Lighting, TV type	Insulation of refrigerator	
Energy services							
Energy services	Road transportation	Heating	Heating or cooling	Steelmaking	Illumination, communication	Food storage	

These are examples of forms of energy and their conversion. See chapter 2, 'Energy flows through systems'. Adapted from UNDP, *World Energy Assessment 2000*, p. 176

Figure 3 Energy conversion steps

the input-output ratios of conversion devices or areas of the economy. Raising energy efficiency implies reducing the *energy intensity* of a process.

The definitions above owe much to environmentalists, who from the 1970s argued that reducing conversion losses was more rational than increasing fossil fuel and nuclear energy supply. They challenged old analytical methods that focused on supply, and that treated consumption as a given, determined by population and economic growth rates. The concept of 'energy services' was coined in 1972 by the researcher Warren Devine, as an alternative to accounting methods that treated renewable energy sources only as substitutes for fossil fuels. He focused on what energy was actually used for, as part of an argument in favour of solar technology.[3]

The environmentalists targeted conversion losses as a key defect of energy systems. 'A rational energy system can virtually eliminate conversion and distribution losses that rob us of delivered end-use energy,' Amory Lovins, a prominent environmentalist critic of US energy policy, said in testimony to Congress in 1976. He advocated 'thermal insulation, more efficient car engines, heat recovery in industrial processes and cogeneration of electricity as a by-product of process heat'. The scale of electricity generation capacity should be matched to network requirements; conservation is 'far cheaper than increasing supply'.[4]

A related issue was the potential for using energy differently – which hinged, again, on the concept of 'energy services'. 'It is the tendency to confuse delivered energy [useful energy, in Figure 3] with needed energy [energy services] that encourages most scientists and engineers to assume that the potential for energy conservation is smaller than it really is,' the energy researcher Robert Ayres argued in 1989. The USA could increase its energy efficiency ten times over 'without exceeding efficiency levels currently claimed for internal combustion engines', he added.[5] Lovins, in his classic proposal for a 'soft energy path', decried the egregious use of types of final energy completely unsuited to the energy service required:

> Where we want to create temperature differences of tens of degrees, we should meet the need with sources whose potential is tens or hundreds of degrees, not with a flame temperature of thousands or a nuclear reaction temperature equivalent to trillions – like cutting butter with a chainsaw.[6]

The history of fossil fuel consumption is partly the history of conversion losses that were reduced, and others that were not. Many disputes have centred on which losses could be reduced with current technology, and at how much cost to whom.

Production and transformation technologies

Fossil fuel production, that is, the extraction, transportation and processing of coal, oil and gas, has become highly mechanised since 1950. These are among the least labour-intensive industries. Consequently, corporations that produce fuels have been able to deliver them to markets at relatively low prices, and in turn have encouraged energy-consuming businesses to invest in energy-intensive technologies and make energy-intensive choices.

Coal production has been characterised by two big trends: mechanisation of underground mining, and increased surface recovery. Productivity rose steadily throughout the twentieth century (in the USA, from 1 tonne per work shift in 1900 to 3.5 tonnes per labour *hour* in 2003).[7] Oil and gas development and production costs have remained relatively low, but have risen steadily over the decades, as production has moved from easier-to-access resources (e.g. large, rich deposits in the Middle East) to more difficult ones (e.g. deep sea or Arctic oil fields).

Technological innovation can complicate the trend for costs to rise, and even reverse it, as it did in the 2000s, when hydraulic fracturing (fracking) and horizontal drilling were successfully combined, enabling producers to access 'unconventional' oil and gas resources, such as those trapped in shale rock. (*Costs* are only one factor in determining *prices*. Since the 1970s, costs have often comprised less than half of oil prices.[8]) Fossil fuel transportation costs – for moving oil and gas by pipelines, coal and oil by rail or sea, and, more recently, liquefied natural gas (LNG) by sea – have also been relatively low historically, encouraging international trade.

Transformation technologies, in the first place for generating and distributing electricity, have played a crucial role in shaping global fossil fuel consumption. The proportion of global fossil fuel consumption accounted for by electricity plants has risen constantly, from 1 per cent in 1900 and 10 per cent in 1950, to 33 per cent in 2011.[9]

Electricity generation from fossil fuels is inherently inefficient. Typically, the fuel is burned to heat water in a boiler, producing steam that drives a set of turbines; the steam is then cooled to water in a condenser and pumped back to the boiler. Power plants' efficiency is limited by the physical principles governing energy conversion,[10] heat loss from the boiler, and the need for energy to run pumps and other systems. Other problems include the way that high-temperature steam corrodes turbine blades and other parts; longer lasting steels have helped. In the USA, average thermal power station efficiency was 24 per cent in 1950, twice what it had been in 1925. In rich countries generally, efficiency rose to around 33 per cent in the early 1960s, with a few individual stations reaching 40–42 per cent. Efficiencies were lower in developing countries: the average in India in the 1970s was 22 per cent. Average efficiencies improved slowly up to the 2000s. In 2005, global average efficiencies were estimated at 34 per cent for coal-fired generation, 37 per cent for oil and 40 per cent for gas; national efficiencies ranged from 28 per cent in India to 45 per cent in Italy.[11]

In the mid and late twentieth century, apart from incremental improvements, two essentially new ways were found to produce electricity more efficiently: combined-cycle gas turbines (CCGT), and cogeneration. CCGT used gas turbines invented in the 1930s for use in aeroplanes, which burn a mixture of fuel and air to drive the turbine blades and expel exhaust gases at very high temperatures. In the 1980s, American and Japanese engineers found ways to use these gases to heat steam for supplementary turbines; this raised power station efficiency to 55–60 per cent.[12] CCGT has the additional advantage of being able to switch on and off much more rapidly than a conventional power station.

Cogeneration, or combined heat and power (CHP), takes the relatively low-temperature (under 100°C) 'waste' heat from electricity production, and distributes it to industry and households. There are technological problems to be overcome – in particular, transporting heat is difficult, and often impossible over longer distances – but, nevertheless, well-planned systems can raise the energy efficiency of fuel combustion to 60–80 per cent (e.g. 30 per cent as electricity, 50 per cent as useful heat). In the 2010s, the world average efficiency of CHP was 58 per cent, compared with 37 per cent for conventional power generation.[13]

CHP is a good example of a technology constrained by systems of power and wealth. Where government, industry and urban planners found ways of using the heat, it flourished. In the Soviet Union and much of Eastern Europe, CHP was installed as standard in the 1950s and 1960s, supplying industry and district heating systems; by 1975 CHP provided 42 per cent of the Soviet Union's urban heat. In the 1970s CHP-supplied district heating systems were also widespread in Denmark, Sweden and Finland; in the 2000s they continued to account for 29 per cent, 49 per cent and 55 per cent of total heat demand respectively in those countries. But where electricity markets were commodified, and shares of the generation market fought for by corporations, CHP declined: in the USA, CHP plants linked to factories generated 30 per cent of electricity in 1930; this proportion fell to 15 per cent in 1950 and 4 per cent in 1975.[14]

During the post-war boom, the US electricity companies, which generated electricity and owned transmission and distribution networks, saw factory-linked CHP as unwelcome competition. They blocked CHPs from feeding electricity to the grid, by refusing to buy it, charging steep prices for the back-up power needed for technical reasons, and so on. Engineers protested that 'up to half' of the country's demand for low-temperature heat, including space heating for homes, could be supplied by CHP. Environmentalists raged at the use in homes of electrical heating, instead of 'waste' heat from electricity production. The 1978 US public utility regulation law removed some of the most egregious obstructions to CHP. (See p. 103.) But the profoundly wasteful use of electricity to produce low-temperature heat remains a central feature of many rich-country energy systems. In the 2000s, when electricity generation firms in former Soviet countries were privatised, governments introduced market pricing of electricity, but not heat. There was no incentive to modernise the old and deteriorating, district heating systems; CHP gave way to boilers on one hand and electricity plants on the other; households and factories opted for autonomous heat sources; and the aggregate energy efficiency of district heating fell.[15]

Electricity systems' efficiency depends not only on power generation, but also on the network of wires used for transmission (high voltage over longer distances) and local distribution (low voltage). In the rich countries, for 40 years up to the 1980s, electricity corporations focused efforts on building larger and larger power stations. Networks were centralised around these plants. In the USA, before the Second World War, power stations typically served a building, a neighbourhood or a town; by 1980 they delivered electricity over distances averaging more than 300 km. Environmentalists argued that economies of scale in generation were married to diseconomies of scale in networks. Transmission and distribution costs of electricity together were more than twice generation costs, they pointed out.[16] Since rural systems were cross-subsidised by urban ones, the point at which decentralisation would have been more economically effective remained blurred.

In the 1980s, the diffusion internationally of CCGT, which can be more effectively deployed in smaller-scale plants, began to reverse the bigger-the-better trend. In public discussion, the virtues of centralisation were no longer taken for granted. (See p. 143.) Big, centralised generation 'has become so important, and so dominates our thinking, that we have long tended to discount the many alternative forms of electricity generation that are smaller in scale and less centralised', Walt Patterson, an electricity researcher who participated in those discussions, argued.[17]

The technologies to which electricity is distributed – lamps, motors, heaters, electronics, and so on – are mostly thousands or millions of times smaller than central-station generators. There are a few larger exceptions, such as aluminium smelting equipment. This mismatch in scale between generation and loads requires the network to divide up generators' output into flows appropriate to the loads – that is, to distribute the electricity. Alternatively, generation itself could be distributed, closer to loads in both location and scale. Technologies for this purpose – distributed (i.e. decentralised) generation – have only in the 2000s begun to be deployed on a noticeable scale.

Energy-consuming technologies

Energy consumption since 1950 has been dominated by the technologies of the second Industrial Revolution. In *transport*, most consumption has been by privately owned cars with ICEs. Electric cars, which are quieter and less polluting, were marketed in the early twentieth century, especially to women, and in 1900 were more widespread in Paris and New York than cars with ICEs. Up to 1930, citizens of many densely populated American and European cities were more likely to travel by electrically powered mass transit than by car.

The petrol car's triumph over these competitors was due not only to social and economic factors but also to technological ones. Electric cars waited in vain for more effective battery storage to be invented, and they could not match the petrol car for speed, range or easy refuelling. Some of these problems have been resolved, and electric cars have reappeared in the twenty-first century, but meanwhile cars with ICEs achieved complete dominance. Advances in engine technology improved cars' average fuel efficiency by a factor of three between the 1970s and the 2010s – but manufacturers lobbied successfully to constrain legal fuel efficiency minima, and used the improved technology not to reduce fuel consumption but to build larger, heavier and faster vehicles.[18]

The gas turbine – a mid-twentieth century invention – transformed air transport, as well as electricity generation (see above) and some industrial processes.[19] Jet engines for aeroplanes were first used during the Second World War, improved upon in the 1950s, and became widespread in passenger planes in the 1960s.

In *industry and construction*, energy-intensive methods of producing materials have been the largest consumers of fossil fuels since the second Industrial Revolution. First among these is iron and steel making. The efficiency of iron smelting (the first stage of steel production) has gradually risen, In 1900 it took 1.3 tonnes of coke (distilled from coal) to produce a tonne of hot metal; by 2000 it took only half a tonne. Nevertheless, ever-greater quantities of coal have gone into steel making. Most steel is produced today in updated versions of nineteenth century plants – open-hearth furnaces, Bessemer converters or more modern basic oxygen furnaces. There has been one truly breakthrough technology: the electric-arc furnace (EAF), which heats raw material with an electric arc rather than directly. EAFs were used to produce alloys during the Second World War, but only widely adopted in the 1980s. They gave steel firms two new flexibilities: EAFs suited mini-mills, that required much smaller capital sums to build; and they could use steel scrap as raw material, thereby bypassing the most energy-intensive stages of steel making, coke making and smelting.[20]

The second most significant metal product, aluminium, is also usually produced with energy-intensive technology developed in the second Industrial Revolution (the Hall-Heroult process). Cement making is another big fossil fuel guzzler, but its efficiency has improved in recent decades, mainly by substituting the old wet process by the newer dry process. In Europe in the 1980s, energy consumption per tonne of cement was 119–160 kilograms of oil equivalent (koe)/tonne using the wet process, and 72–95 koe/tonne with the dry process. By the 2000s, most countries were achieving average energy consumption of 76–81 koe/tonne of cement, with the USA, Canada and China lagging behind at 100–110 koe/tonne.[21] The big early twentieth-century innovation in materials production was petrochemicals (chemicals derived from hydrocarbons). Among its products are explosives; synthetic fibres; and cheap plastics, used in packaging, construction and electronics, consumption of which has doubled once every 15 years since 1950.[22]

A key turning point for *agriculture* was the discovery in 1913 of the Haber-Bosch process that synthesised ammonia, out of nitrogen (from the air) and hydrogen (from fossil fuels, mainly natural gas but also oil and coal). The process was first used to make explosives during the First World War and then to make fertilisers. In the second half of the twentieth century, ammonia production became more efficient: the energy used per tonne of ammonia approximately halved.[23] Industrial agriculture emerged as a system, using chemical fertilisers, mechanical (usually fossil-fuel-powered) farm equipment and high-yield grains and genetic engineering. No wonder it was defined as 'the use of land to convert petroleum into food.'[24] These methods, developed in the USA in the post-war years, were exported to developing countries during the so-called 'green revolution' of the 1970s, and continue today to play a central role in shaping the way that food is produced and consumed.

The spread of electricity networks, from the second Industrial Revolution onwards, paved the way for new electrical technologies. In the first half of the twentieth century, electric motors came into widespread use in manufacturing. Electrical and electronic household appliances, already common in some rich countries in the first half of the twentieth century, proliferated everywhere in the second half. Rich-country households first acquired radios, cookers, fridges and washing machines, and then moved on to TVs and electric fans, all of which transformed how people lived and, in particular, the way that housework was done. From the 1980s, the third technological revolution brought personal computers, mobile phones and an array of other gadgets into households.

Technological means of reducing fuel consumption

The industrial boom of the 1950s–1960s in the rich countries was characterised by the availability of cheap fossil fuels and the proliferation of energy-intensive, and even wasteful, types of consumption. In the 1970s, the sharp rise in fuel prices spurred research into changing technological processes, particularly in industry, to reduce fuel consumption. The engineers knew what they were looking for. In 1973, research commissioned by the US state of Illinois showed huge potential energy savings from all stages of car manufacture (particularly steel and aluminium making), from recycling, and by making cars to last 30 years instead of 10. (See p. 84.) In 1974 the US government commissioned a project that quantified the potential for using solar, instead of fossil energy, to provide process heat in industry where only lower temperatures were needed. Hardly any such potential was realised, though. Companies had other priorities, especially once energy prices stabilised, and then fell, in the 1980s. Daniel Spreng, a researcher of energy flows, in 1988 blamed 'the overzealous belief in growth [that] leads directly to a large waste of resources', and specifically to the construction of unneeded production capacity, and wasteful uses of machinery, in industry.[25]

Modern buildings, often built energy-intensively, which then use energy wastefully for space heating and air conditioning, have long been a target for energy conservation advocates. The technological potentials for energy conservation have hardly ever been realised. Amory Lovins argued in 1992 that 'the institutional framework within which buildings are financed, designed, constructed and operated' – including developers' focus on minimising capital costs, and owners' indifference to reducing heating costs – was to blame.[26] Since then regulation has improved only in a handful of countries.

Research into technological means of reducing fuel consumption received renewed impetus from the discovery of global warming and resulting political processes. A good example is research of the production of steel, aluminium and other materials, conducted in the 2010s at Cambridge University. In the case of steel, the researchers found that, of the scrap fed back into steel mills, less than half was end-of-life scrap (objects made from steel but no longer used); most was material left behind by steel making and processing. A 'never-ending [and extremely wasteful] internal melting loop of 360 mt of steel' was maintained (meaning that amount of steel never exited the production process). Losses during fabrication (when steel is turned into shapes needed by manufacturers and construction companies) could be slashed; more goods could be made with less energy-intensive long products (e.g. bars, beams or wire), instead of sheet steel; and construction, in particular, could use designs that require less steel. The reasons companies made materials-wasteful decisions included the priority given to reducing labour time, and 'habit'. The losses in aluminium production are even more eye watering: 'around half of all liquid aluminium produced each year never reaches a final product', and recycling would slash energy use.[27]

The relative costs of materials and labour discouraged producers from reducing materials output. 'Using less materials incurs other costs, such as higher labour costs, higher tooling costs or higher costs for higher performance materials'; if companies were to be motivated by government to take material efficiency options, they would want to save more than 75 kilograms of steel or 20 kilograms of aluminium for an hour's labour.[28] The researchers concluded that total potential energy savings from 'practically achievable design changes' to the most energy-hungry technological systems – building design, vehicles and industrial systems – amounted to 73 per cent of global primary energy use. The barriers to energy demand reduction were primarily 'economic and socio-cultural'.[29]

Non-fossil energy producing technologies

Higher fossil fuel prices in the 1970s focused governments' attention on other ways of producing energy, as well as on conservation. Nuclear power, which fits both with military strategies and with centralised electricity generation, always had the lion's share of political support, and was widely expected to become the main alternative to fossil fuels. Hydro power expanded, particularly outside the rich world. Research funds went into solar and wind power technologies, and heat pumps, but – as with energy conservation – much of the progress made was reversed in the 1980s. By the 2000s, global warming put on the agenda the issue of a transition to a non-fossil-fuel energy system.

State funding for research and development is a good measure of the relative importance given to nuclear and renewable technologies. Between 1950 and 1973, OECD countries spent more than $150 billion on nuclear and practically nothing on renewable energy sources. Between 1974 and 1992, these countries spent $168 billion on nuclear research and $22 billion on renewables.[30] In addition to research funds, nuclear power received economic subsidies (e.g.

tax breaks and favourable electricity purchase rates). Nevertheless, it was constrained by social factors – mainly, opposition on the grounds of its relationship to nuclear weapons, and safety issues.

The most important solar technology, photovoltaic (PV) cells made of semiconductor materials that convert sunlight into electricity, was devised in 1952 by scientists seeking new sources of electricity for telecommunications. Their first cell was 6 per cent efficient (converted 6 per cent of the light it received into electricity); twenty-first-century cells are 10–25 per cent efficient. Concentrator photovoltaic cells, which achieve efficiencies of more than 40 per cent, are still in development.[31] For PV systems to work at scale, though, other problems had to be solved – and by the 2000s had begun to be solved – including developing good batteries, physical support structures for panels and inverters to feed electricity grids, and sourcing raw materials. All these factors influence up-front capital costs, that were historically high, and – since solar largely lacked nuclear-type state support – frustrated efforts to compete with incumbent fossil fuel systems. Nevertheless, capital costs for solar PV fell from more than $76/watt of capacity in 1977 to less than $1 in 2010, and are still falling.[32]

In the USA, state aid to solar rose from a 1960s average of $500,000/year to $500 million in 1979, resulting in a mini boom. A large chunk of subsidies was directed at domestic solar water heaters (a mistake, according to energy specialists); by 1980, this business achieved $1 billion annual turnover. A change of government – the replacement as president of Jimmy Carter, who demonstratively installed solar panels on the White House, which were removed by Ronald Reagan – resulted in subsidies being withdrawn. Nearly all the progress made was reversed. It took another 20 years, until the turn of the century, before substantial state support was given to solar PV – by Germany and Japan this time, followed by Spain and then China – paving the way for rapid expansion of solar electricity generation, and a further corresponding fall in costs, in the 2010s.[33] The future development of solar and other renewables will not only have to overcome social and economic problems, but also resolve technological issues mainly related to intermittency – that the sun shines and the wind blows intermittently. (See pp. 185–6.)

Wind turbines, which had been used to generate electricity, particularly in Denmark, since the early twentieth century, benefited from the 1970s oil price shocks more slowly than solar. In the USA, they were diffused only after 1978 legislation that opened the electricity grid to small generators. (See p. 103.) US wind power was concentrated almost entirely in California, which in 1985 produced more than 95 per cent of it. Thereafter, the withdrawal of state support resulted in the collapse of US wind turbine manufacturers, while Danish producers – who, with strong government support had perfected the more efficient three-blade upwind turbine – took a growing share of the world market.[34]

Heat pumps, like solar, boomed in the 1970s but in some countries were pushed back to the margins in the 1980s. There was a different experience, though, in Sweden and Switzerland, where government support (research funds, and favourable tax regimes for commercial providers) enabled the technology to develop, for engineers to learn from each other and make improvements, for costs to fall and for heat pumps to increase their share of energy provision.[35]

Technologies of the third Industrial Revolution

The innovation from the 1980s, and diffusion from the 1990s, of a cluster of information and computing technologies is justly described as a third Industrial Revolution. Starting with the transformative development of the microprocessor, this revolution has generalised digital computers; transmission technologies such as computer networking, the Internet, digital broadcasting and (later) third- and fourth-generation mobile phones; and has ushered in digitally driven manufacturing and new types of robots. Claims that this process would reduce the economy's impact on the environment have not been realised; in some respects, including fossil fuel consumption, it has produced new burdens. All these technologies have spurred large-scale manufacture of ultra-energy-intensive materials, such as silicon wafers, which embody 800 times as much energy per unit of weight as steel (although they are of course used in much smaller quantities).[36] Moreover, the data centres and telecoms infrastructure on which the systems rely wolf down electricity. (See p. 162.)

Even more striking than these burdens, though, is the timidity of efforts to apply information technologies to the problem of reducing fossil fuel consumption. Although the global warming danger was well known when these technologies were being diffused, their potential to remake electricity networks and industrial processes – which have been clear to engineers for many years – have not been realised.

The data gathering and communication functions used every minute by the Internet could manage electricity flows. 'Smart grid' technology, currently used to offer household consumers commercial deals, could do much more; for instance, by coordinating multiple small sources of supply in distributed (i.e. decentralised) generation systems. This would also improve the conditions for technologies such as solar PV to operate at smaller scales (for example, in households as opposed to large solar farms). Yet, as researchers at the Electric Power Research Institute noted, electricity is 'the last industry in the western world to modernise itself' using this technology. The *Global Energy Assessment*'s authors wrote:[37]

> It is indeed a supreme irony that computers, sensors and computational ability have transformed every major industry except power-generation, whose product is the lifeblood of the modern global economy.

Why is there a disconnect between the available technology and its deployment? One thing to consider is:

> Today's bulk-electricity supply and delivery systems were typically designed and built with the primary objective of keeping pace with the rapid growth in demand associated with initial electrification. Those days are long over throughout the developed world, yet little has been done in most countries to update either the infrastructure or the business incentives to focus on efficiency and quality. The electricity meter, for example, still holds retail consumers hostage to an electricity supplier monopoly over which they have essentially no market leverage. Technology is available to break down this iron curtain meter, just as the internet transformed communications. [...] The incentives should be to add maximum value to each electron, not to maximise the quantity of bulk electricity sold to captive consumers.

The technology has leaped forward, but faced social, economic and political barriers. Bulk supply systems and meters that hold customers hostage have obstructed change. Behind them have stood energy corporations' vested interests. The prospect of large numbers of small solar PV electricity suppliers, brought together by smart grids, posed a threat to incumbent firms, two writers on electricity history pointed out. Distributed generation 'brings new stakeholders – millions in the case of PV – into the supply side of the industry, with potentially revolutionary intent'. But as things stand, these small suppliers 'still need the grid more than the grid needs them'. By 2012, both Google (through its PowerMeter subsidiary) and Microsoft (with Microsoft Hohm) sensed an Internet-like business opportunity, entered the market, and exited again.[38] Electricity grids require much more coordination than the Internet does. So far, movement towards decentralisation has been in the teeth of resistance from the incumbent electricity firms.

The theme that emerges, both in terms of the adaptation of existing technologies and the diffusion of new ones, is that economic and political factors have been decisive in determining the pace of change.

3
Energy in society

Changes in the economic framework in which fossil fuels have been supplied are outlined in this chapter; then, economic and social trends that drove increases in consumption; the relationship of population growth and fossil fuel consumption; and the reflection of widening social and economic inequalities in the way fossil fuels were distributed and consumed.

Energy supply in economic history

In the last two centuries, energy has mostly been brought to its consumers in three overlapping but distinct ways: *as a commodity* (e.g. traded oil, gas or coal); *as a state benefit* (e.g. as cheap electricity for industry or urban residents); or *as non-commercial energy* gathered from nature by families or communities (e.g. fuel wood in the countryside of developing countries).[1]

Commodified energy became dominant in the late nineteenth and early twentieth century. Until the eighteenth century, only a small proportion of the energy consumed anywhere was traded. People mostly burned biomass or coal they had gathered themselves or bought locally; the main energy source for agriculture was human and animal muscle power. Townspeople and industrial enterprises were more likely to buy wood and coal from others, but they were a small minority. Coal trading expanded significantly in eighteenth-century Britain; it helped to pave the way for the Industrial Revolution and for urban development; these changes, in turn, drove up demand for coal, primarily in industry. (See pp. 11–13.) Widespread use of electricity, oil and gas began in the late nineteenth century, by which time the capitalist economy was mature and its imperialist expansion was underway. Oil and gas were produced and marketed by corporations from the start. In the case of electricity, both corporations and governments invested in networks, and economically, two types of provision – as a commodity and as a state benefit – clashed and/or combined. Technologically, systems of energy provision became more complex; economically, the provision of energy as a commodity became dominant. David Nye summarised the social consequences of these changes:

> The ability to consume energy without having to produce it is partial and recent. It also has its problematic side. As farms, cities, businesses and

households gave up energy self-sufficiency, they become dependent on distant and often anonymous sources of power, whether coal mines, dams, oil wells or gas fields. Such superior methods of consuming power were essential to getting more work out of labourers, to producing goods more cheaply and to freeing more labour for other tasks. European expansionism and imperialism would have been difficult to sustain without improved power systems.[2]

This rift between energy consumers and producers was further deepened as electricity networks spread. Extended grids, centralised generation, and the aggregation of diverse sources of supply has in recent times 'allow[ed] end-users to disengage from the electricity industry other than through paying their bills', the energy researchers Stephen Healy and Iain MacGill have written. In rich countries, the grid 'appears effortlessly to supply just about any amount of electricity'. This disengagement will be reversed in a future decentralised system, they argue.[3]

By 1950, of the fossil fuels, only a small proportion of coal continued to be produced and consumed outside markets. Even today, more than 400 million people use coal for cooking, mostly in the Chinese countryside; many of these gather it themselves. Not so with oil, the quintessential commodity. The proportion traded across borders has risen even faster than the total produced: in 1950 less than one third of crude oil crossed a border on its way to the consumer; by 2015, more than two thirds.[4]

Ideas about *energy as a state benefit* first emerged as nineteenth century urban populations lost their energy self-sufficiency in the way described by Nye. As electricity became widely available in cities in the late nineteenth century, in the USA it was sold by private business, while in European countries, the state at national or municipal level provided it. (See pp. 20–1.)

Electricity fired the imaginations of socialist opponents of capitalism, who saw its potential as a high-tech platform for an equitable, post-capitalist society. In 1879, August Bebel, a leading figure in the German Social Democrats, the world's largest political party, foresaw electrical networks and appliances sweeping away the burden of housework on women. Charles Steinmetz, the German-born socialist who migrated to the USA and as head of research at GE helped to invent alternating current, believed that a national electric grid would engender technological interdependence that would pave the way for a new social system. The Russian anarchist Petr Kropotkin saw electricity as a means of strengthening worker cooperatives against big corporations.[5] This radical optimism was summed up, at the dawn of the twentieth century, by the French novelist Emile Zola. In his novel *Work*, he had his character Luc Froment, an electrical engineer who endeavoured to create a collectivist community based on Charles Fourier's cooperativist socialism, say:

> The day must come when electricity will belong to everybody, like the water of the rivers and the breezes of the heavens. It will be necessary to give it abundantly to one and all, and to allow men to dispose of it as they choose. It must circulate in our towns like the very blood of social life.[6]

In the twentieth century, when socialist governments began to be formed, they prioritised electricity. Social democrats saw it as an infrastructure service for both capital and labour; the state socialists who ruled the USSR saw it as an essential basis for industrialisation and modernisation. After the Second World War, the success of relatively rapid Soviet electrification influenced developing country governments seeking models to emulate – while western capital's institutions, such as the World Bank, advocated commodified electricity, and electrification financed by capital markets. Electricity's potential as both a commodity and a state benefit made it a battleground between contending social forces.

Energy that changes hands in the economy – both commodified energy and energy provided as a state benefit – are identified in this book as commercial energy.

Non-commercial energy remains dominant for most people in the countryside, and a smaller proportion of city dwellers, outside the rich world. These communities rely mainly on biomass – mostly fuel wood, but also charcoal and animal dung – for cooking and heating, and often on human and animal muscle power for subsistence farming. The absence of electricity and modern fuels cruelly shapes the use of labour. The task of collecting biomass – often involving journeys on foot of many kilometres each day – falls largely on women and children. I call such energy sources non-commercial, rather than 'traditional', which might imply that they have no costs: actually, the cost in terms of labour is high.[7]

There are about 1.2 billion people who have no access to electricity, and 2.7 billion who use biomass to cook: the number living outside the commodified energy system is between these two. (See Table 3, p. 42.) The dividing line between the non-commercial and commercial energy systems is fuzzy. It is close to the dividing line of electrification: in rural areas outside the rich world, electrification requires the monetisation of economic relations, since households need cash income to pay for electricity and electric appliances. But this is not an exact correlation. Some rural families living mainly outside the commercial system sometimes access kerosene or diesel in markets. Some urban families living on the edges of the commercial system fall back on non-commercial biofuels for some important uses. Research in the late 1970s found that poor urban families bought, bartered or collected charcoal and biofuels, as well as buying bottled gas or electricity. The grey area has persisted over time: 30 years later, research of urban households in China and India found that barter exchange was often still the norm. Much also depended on ease of access: the larger the Indian city, the harder it was for poor households to gather their own firewood, and they would then have to buy it.[8] The authors of the *Global Energy Assessment* (2012) estimated that 230 million urban dwellers have neither electricity nor modern fuels; another 470 million or so have some electricity but no modern fuels for cooking.[9]

The history of fossil fuel consumption since 1950 is primarily the history of commodified energy. The provision of energy as a state benefit, which expanded during the post-war boom, was eroded by the neoliberal offensive of the 1980s–1990s but has by no means disappeared. Non-commodified, and non-fossil-fuel, energy remains vital for a large minority of the world's population, including its poorest sections. But in terms of sheer volume, commodified energy supply was dominant in 1950 and has become overwhelmingly dominant since.

Social and economic causes of rising consumption

The six social and economic trends that jointly account for most fossil fuel consumption growth since 1950 – and have been manifested through the technologies described in Chapter 2 – are: electrification, industrialisation, the transformation of the labour process, urbanisation, motorisation and the growth of material consumption and consumerism.

Electrification began before the Second World War in rich countries, when networks were installed. Consumption expanded sharply in the post-war boom. In the rest of the world, electricity was available only in a few urban islands in 1950; today, networks reach most urban areas and large parts of the countryside.

Industrialisation was also concentrated in the rich countries in the first decades after the Second World War. Energy consumption by industry expanded steadily until the 1970s. The recessions of the 1970s pushed energy use by rich-world industry down, and in 1983 it was one-fifth lower than it had been a decade earlier.[10] Then it rose again, notwithstanding efficiency improvements. Even in the 1980s, roughly 90 per cent of the world's industry was sited in rich countries. But as production processes – and in particular the energy-intensive manufacture of raw materials (metals, cement and fertilisers) – were relocated to developing nations, the total use of energy by industry kept rising. Between 1970 and 2005, annual global output of cement rose by 271 per cent; of aluminium by 223 per cent; of steel by 84 per cent; and of ammonia (the fossil-fuel-based ingredient for fertilisers) by 200 per cent.[11] Between 1971 and 2011 total energy consumption by industry rose by four-fifths; other sectors such as transport increased their energy consumption more rapidly, though, so industry use fell as a proportion of the total, from 33 per cent to 28 per cent.[12]

The technological changes in industry, and industrialised agriculture, made possible by fossil fuels were associated with *changes in the labour process*: the continuation of substitution for hard physical labour previously done by people

Table 3 People without access to electricity, and people relying on biomass for cooking

In millions, and % of the population	1970 People with no electricity		1990 People with no electricity		2000 People with no electricity		2000 People using biomass to cook		2009 People with no electricity		2009 People using biomass to cook		2013 People with no electricity		2013 People using biomass to cook	
	m	%	m	%	m	%	m	%	m	%	m	%	m	%	m	%
Africa	315	86%	473	75%	522	66%	583	72%	587	58%	657	67%	635	57%	754	68%
Sub-Saharan Africa	258	91%	412	84%	497	77%	575	89%	585	69%	653	80%	634	68%	753	80%
Developing Asia	n/a		n/a		1041	36%	n/a		799	22%	1937	55%	526	14%	1895	51%
China	n/a		n/a		n/a		706	56%	8	1%	423	32%	1		450	33%
East Asia/China	685	70%	603	44%	196	13%	n/a		n/a		n/a		n/a		n/a	
India	n/a		n/a		n/a		585	58%	404	34%	855	75%	237	19%	841	67%
South Asia	616	83%	809	68%	860	59%	n/a		n/a		n/a		n/a		n/a	
Latin America	158	55%	134	30%	55.8	13%	96	23%	31	7%	85	18%	22	5%	65	14%
Developing countries	n/a		n/a	54%	1634	36%	2390	52%	1438	27%	2679	54%	1200	22%	2722	50%
World	1877	51%	2124	40%	1645	27%	n/a		1441	21%	2679	40%	1201	17%	2722	38%
World total population	3682		5309		6127				6764				7182			

Sources: IEA, *Energy Poverty: how to make modern access universal* (2010); IEA, *World Energy Outlook* (2002 and 2015); UN Population Division, *World population prospects*, 2015 revision

and animals, and the substitution of, and changes in, other types of labour by electrical appliances. The fossil-fuel-dominated economy also redefined the division between industrial, agricultural and domestic labour – for example, giving rise to the manufacture of food, where previously that food would have been made at home. Moreover, it transformed the sphere of domestic labour, relieving some burdens but creating others. (See pp. 90–2.)

Changes in the labour process began, during the post-war boom, to produce an unprecedented flood of affordable commodities in rich countries. Cars and household items previously supplied to a wealthy minority became accessible by most people. Automated assembly lines drove down costs; manufacturers then focused on design variation, marketing and new products to encourage replacement purchasing. From the 1980s, especially with the advance of computer and other technologies, the throwaway culture, in which it was cheaper to replace a product than fix it, expanded.[13] The other side of this coin was the trend towards rich-country workers working longer hours, for higher pay, and spending it on time-saving items, such as travel by private car, pre-prepared food or household appliances. (See pp. 126–7.) In the twenty-first century, new technologies made possible just-in-time manufacture and globalised delivery systems. All this made manufacturing, commerce and personal consumption more fossil fuel intensive.

Urbanisation tends to increase fossil fuel consumption per head, but not straightforwardly. Industry, transport, buildings and other infrastructures weigh the scales; so do amenities such as electricity, heat, hot water and gas, whatever gaps there are in provision. A pioneering study of energy consumption published in 1981 found that the geographical intensity of consumption in rich-world cities was generally around 5 Watts/square metre (3.8 million toe/year/square km). The authors advanced the hypothesis that, given the role of technological systems in determining consumption levels, and the greater population densities in poorer cities, urban areas outside the rich world might have similar levels of energy consumption per square kilometre. Subsequent research showed that they do not. Inequalities between cities, and the proliferation of private car transport and suburban dwellings in the Global North, made cities there substantially more fuel-intensive, a UN report stated in 2008. Energy flows in the richest megacities (those with more than 10 million population) were found to be 28 times greater per head than in the poorest.[14] The authors of the *Global Energy Assessment* (2012) estimated energy consumption by the urban population (then a little more than half the world total) at 60–80 per cent, with a central estimate of 75 per cent, using production accounting, and nearer to 80 per cent, using consumption-based accounting. (On accounting methods, see Appendix 1.) The urban share of high-quality energy carriers such as electricity, diesel oil and petrol, is higher still.[15]

Urbanisation is a jagged and uneven process, very different in the Global South in the late twentieth and early twenty-first centuries from in Europe and North America a century earlier. Sheer scale is one factor. About 14 per cent of the world population was urban in 1900, 30 per cent in 1950, and 54 per cent by 2015. Between 1950 and 2005, while the world population grew from about 2.5 billion to about 6.5 billion, and the rich-world urban population went up from 340 million to 730 million, the urban population of Asia rose from 175 million to 1.3 billion, of the Middle East and Africa more than tenfold from 44 million to 478 million, and of Latin America from 68 million to 429 million. In 2010–15, 77 million people were added to the urban population each year, compared to 57 million in the 1990s. The number of megacities rose from 14 in 1995 to 29 in 2015.[16]

Whereas urbanisation in the rich world was almost always associated with industrialisation, pulling labour into factories, the developing-country urbanisation of the last half-century has brought hundreds of millions of people into cities to eke out a precarious existence with no work or irregular work. The population of slums has expanded much faster than urban population in general. In 2001 the UN counted 924 million slum-dwellers, comprising 32 per cent of the world's urban population (59 per cent in south-central Asia and 72 per cent in sub-Saharan Africa).[17] Urban fossil fuel consumption is deeply unequal. 'Urban sprawl and motorisation come hand in hand with the expansion of slums and gated communities, and the associated social divide', UN researchers found, with 'the better-off classes producing the bulk of the emissions'.[18]

Motorisation is the fourth big social and economic trend that has driven fossil fuel consumption growth. Consumption of energy for road transport, all but a minute proportion in oil-based fuels, tripled in the 40 years between 1971 and 2011. Most of this fuel was consumed by cars, predominantly privately owned cars in rich countries. Rich-country car ownership rates soared in the last third of the twentieth century: in 1973 there was one car between nine people in Japan and one between two people in the US; by 1998 all OECD countries had more than one car for every three people; the USA had more than two.[19]

Peter Newman and Jeffrey Kenworthy, summarising decades of research, argued that 'automobile dependence', particularly in cities, is a prime cause of high fossil fuel consumption. In the (mostly rich-world) cities they surveyed, the number of vehicle kilometres travelled per head of population rose by 42 per cent in the decade to 1970, 26 per cent in the decade to 1980 and 23 per cent in the decade to 1990. After that, growth slowed to about one third of those rates, suggesting it had reached saturation level. Low urban densities (i.e. urban sprawl, with people living in larger houses, further from their workplaces) and the retreat of public transit services were correlated with growing dependence on cars.[20] The supply of urban highways, which 'directly facilitate greater car use and energy use in cities', is crucial. Road construction and parking provision have historically provided a gigantic subsidy to car drivers at others' expense. Decades after urban cars' contribution to global warming was well understood, city authorities continued to require construction firms to build parking spaces in city centres – 'a subsidy for the wealthy paid by the poor' that further undermines public transport and cycling, as the *Economist* noted.[21]

Atlanta in the USA has become the symbol of car-dependent living: its transport-related carbon emissions are eleven times higher per head than those of Barcelona, Spain, which has similar population and GDP per capita (and 100 times higher than those of Ho Chi Minh City, Vietnam). The greatest distance between two points in Atlanta's city area is 137 km, compared to 37 km in Barcelona; the proportion of trips made on foot is 20 per cent in Barcelona; in Atlanta it is too small to be recorded.[22] 'In sprawled metropolitan areas, people use their cars to do everything,' an economists' study of US transport found; in the most spread out cities, even 78 per cent of trips of 1.6 km or less are made by car. The current model of urbanisation, a UN report concluded in 2016, 'engenders low-density suburbanisation – largely steered by private, rather than public interest, and partly facilitated by dependence on car ownership. It is energy intensive and contributes dangerously to climate change.'[23]

The final trend to be considered is the growth of *household material consumption* (of goods and services generally, not only the direct consumption of energy), and the *consumerism* that goes with it. In the twentieth century, material consumption, above – and sometimes far above – the subsistence level has spread, from social elites numbering a few thousands, to tens and hundreds of millions of people in rich countries. From the 1980s it spread to smaller proportions of middle-class people outside the rich world too.

Such material consumption affects fossil fuel consumption levels in two ways. First, indirect consumption, when consumers purchase cars, houses, and other goods, the production of which involves fossil fuel use. Second, direct household consumption of energy, as electricity; heat; coal and gas for heating and cooking; and so on. This direct consumption doubled between 1971 and 2011, staying consistently at 23–4 per cent of global final energy consumption. (See Table 5-b, p. 60.) Even today, it is predominantly a rich-world phenomenon; biofuels remain the largest household energy source outside the rich world.

A part of this direct household energy consumption substitutes for and reduces the burden of domestic labour, mainly through the use of electrical appliances such as refrigerators, cookers, washing machines and vacuum cleaners. There is a distinction between this energy consumption associated with domestic labour and consumption for communication or leisure, such as for TVs and radios.[24] There is no neat divide. Personal computers might be used for paid-for labour, domestic labour, or for fun; and the same refrigerator might contain bottles of wine for leisure and labour saving food supplies.

Nevertheless, there is a difference. A washing machine is as central to the labour process for a rich-world urban parent as an electric drill is for a joiner, but in both their households, Game Boys or snowmobiles are leisure equipment. In rich countries, communication and leisure technologies have over the last 40 years become significant consumers of electricity.

There is a further distinction between *discretionary* consumption (of energy used as a direct result of households' own decisions), and *non-discretionary* consumption (largely determined *outside* the household, by economic, social and technological systems and the people that control them). All too often, researchers assume, or imply, that the level of consumption is chiefly determined by households' decisions. And certainly, many rich-world households could tomorrow slash their energy consumption without any significant suffering or inconvenience. But focusing on that hypothetical, while ignoring the ways that households' consumption has increased *due to the actions and decisions of others*, produces a distorted picture.

This is not a new insight. More than 40 years ago, researchers commissioned by the US government's National Research Council dismissed the idea that 'consumers are basically free to choose among actions that imply different amounts of energy use'. Consumers' choices, they argued, were limited (i) by intermediaries, such as purchasers of buildings and energy-using equipment; organisations such as utility companies who are incentivised to distribute energy less efficiently; and engineers, designers, architects, financial institutions and standard setting organisations who shape a building's energy consumption before its occupants move in (e.g. by assuming that all rooms had to be heated); (ii) by makers of consumer products who determine levels of indirect consumption; (iii) by long-lived capital stock (e.g. housing) that locks in patterns of consumption; and (iv) by limited access to capital for energy conservation equipment.[25]

To this list could be added the effects of government policy, which, in rich countries during the post-war boom, kept taxes on fuels low, and used cheap energy as a subsidy both to households and industry. This political approach, which prioritised social control over energy efficiency, persisted after the dangers of global warming became known in the 1980s. Moreover, energy is used by corporate and public infrastructure, from supermarket supply chains to night-time urban lighting, over which individual consumers have no control. In the 1990s, environmental sociologists researched these issues. David Goldblatt argued against the 'conventional individualistic, behavioural focus' of energy consumption research, and advocated a 'social-revealing approach' that understands that energy consumption is embedded in products, services and systems and distinguishes its discretionary and non-discretionary forms.[26] In this book I endeavour to do that.

Fossil fuel consumption and population

The view that fossil fuels are consumed by and through technological, social and economic systems, and that consumption growth is driven mainly by social and economic changes, is presented in this book as a counter-weight to the oft-repeated and misleading generality, that population growth is one of the main drivers. In fact the relationship between population growth and fossil fuel consumption growth is complex and heavily mediated through systems.

Rich, heavily industrialised and urbanised countries with small populations consume more fossil fuels than large countries in the Global South with big, mainly rural, populations: Belgium has one citizen for every 14 in Bangladesh, but consumes two-thirds more fossil fuels.[27] As for population *growth*, in the Global South, the correlation with fuel consumption is very weak; in the north, since the 1970s, it has been associated with less rapid fossil fuel consumption growth than previously, due mainly to changing economic structure. Sub-Saharan Africa accounted for 18.5 per cent of global population growth in 1950–1980, and 10.7 per cent in 1980–2005, but only 2.2 per cent and 2.4 per cent respectively of global CO_2 emissions growth (which is mainly caused by fossil fuel consumption) during those periods. North America accounted for 4 per cent of population growth in both periods, but for 20 per cent of CO_2 emissions growth in 1950–1980, and 14 per cent of it in 1980–2005. There are of course many indirect relationships between population growth in the Global South (where it is concentrated) and fuel consumption growth: for example, the expansion in the 1970s of fossil-fuel-intensive agriculture supported more rapid population growth. But statistically these are masked by the very low levels of direct and indirect personal fuel consumption in the Global South. Researchers have found equally weak correlation between population growth and other types of consumption.[28]

Figure 4 shows the relationship between population and fuel consumption trends for several large countries. In India, fossil fuel use rose *at a slower rate* than population for much of the period since 1980. Moreover, most incremental fuel use was by industry and relatively small urban populations. Between 1981 and 2011, improvements in electricity access that pushed up household access levels from around 20 per cent to around 70 per cent, and provided electricity to 650 million mainly rural people for the first time, accounted for just 3–4 per cent of the growth in the country's greenhouse gas emissions. The total direct and indirect use of electricity by those 650 million people accounted for 11–25 per cent of emissions growth; most of the rest was due to urban populations and industry.[29] In China, the acceleration in fossil fuel consumption growth from 2001 was clearly caused by rapid changes in the economy. If a demographic trend is relevant, it is the large-scale movement into towns in the 1990s and 2000s, not the actual level of population. This urbanisation accounted for an

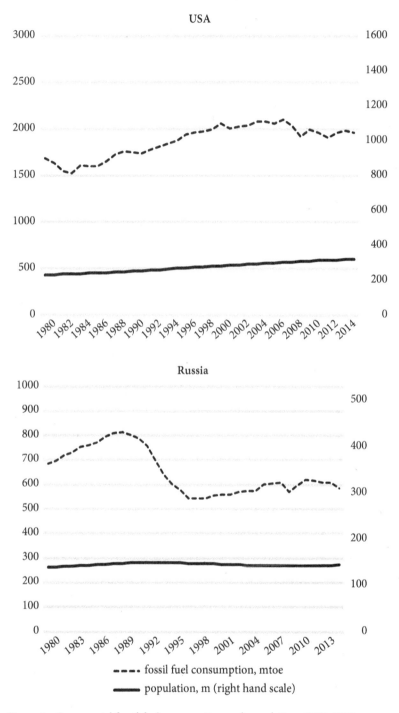

Figure 4 Commercial fossil fuel consumption and population, 1980–2015

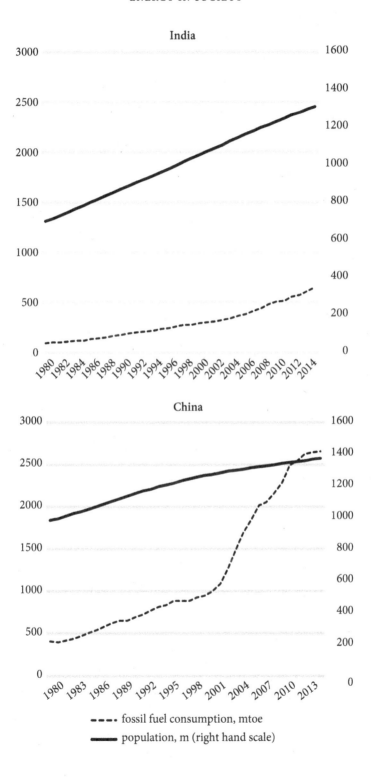

increase in household consumption, but was dwarfed by coal consumption in industry, and coal consumption for electricity used mainly by industry.

The complex relationship between population growth and fuel consumption growth is further illustrated by the graphs for the USA and Russia in Figure 4. The USA's fuel consumption remained since 1980 between three and ten times greater than India's, despite the population remaining between two and four times smaller. The changes in the fuel consumption level are related to economic shifts: it fell in the early 1980s, and again in the late 2000s, due to economic recessions. It grew at all other times, finally levelling out in the 2010s, thanks mainly to efficiency gains. In Russia, population fell very gradually from the late 1980s, in contrast to most countries. But the crash of fossil fuel consumption in the early 1990s was caused by an unprecedented economic slump. From the mid-1990s as the economy slowly recovered, so did fuel consumption – although the population continued to shrink.

The conclusion is not that there is no relationship at all between population growth and fossil fuel consumption, but that there is no *direct causal* relationship. The relationship is complex and mediated. People consume fossil fuels through technological and social systems, and the predominant factors that determine consumption levels are the ways that those systems evolve.

A mirror of inequality

The widening gap between rich and poor, a central feature of recent economic history, is reflected in the level of fossil fuel consumption, and how the fuels are consumed. At one extreme is the high level of consumption by industries, infrastructure and consumers in the rich world; at the other, the absence of commercial energy systems from swathes of the countryside in poor countries. The differences in the *way* energy is used are essential. Rich countries support high living standards by providing food, water and energy through systems that depend on high inputs of energy, capital and materials. In developing countries, many people lack the materials and energy needed to take them beyond the most basic levels of subsistence.

One measure of inequalities, albeit imperfect, is per capita energy consumption. In the 1960s, it was estimated to be 40–50 times higher in North America than in Africa and developing Asia. In the 1980s, it was estimated to be 60–70 times higher in the USA than in Bangladesh. In 2007, Vaclav Smil estimated that in Canada it was 450 times higher than in the poorest African countries – but that, since the poorest people in the poorest countries do not directly consume any fossil energies or electricity at all, the gap between them and the average Canadian 'may easily be larger than 1000 fold'.[30]

National per-head consumption statistics are shown in Figure 5. Crucially, they do not reflect inequalities within nations. In India, the energy researcher

Shonali Pachauri found that between 1983 and 2000, while the number of people counted as suffering energy poverty fell and tens of millions gained electricity access, the proportions of energy consumed by the richest and the poorest scarcely changed. (See p. 130.) The economists Lucien Chancel and Thomas Piketty attempted to compare trends in global inequality, in terms of CO_2 emissions, between countries and within countries, in the 15 years prior to the Paris climate conference in 2015. They concluded that, overall, global inequalities between individuals had abated, due to the rise of wealthy upper and middle classes in some developing countries and the stagnation of incomes in rich countries – but that inequality *within* countries had widened.[31]

Differences in economic structure and infrastructure are also invisible in consumption-per-head statistics. For example, Figure 5 shows that Germany's and Russia's consumption per head is in the same range – despite Russians' living standards being lower, and their personal consumption more modest, than

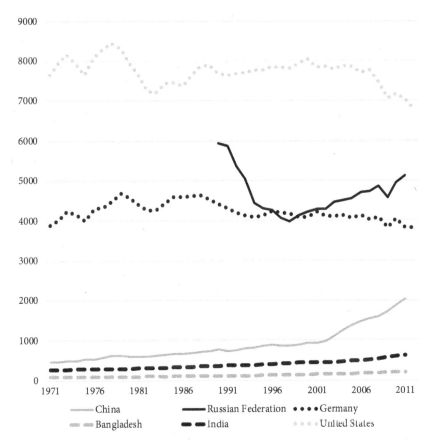

Figure 5   Energy consumption per person per year, 1971–2011
kg of oil equivalent

Germans'. Russia's numbers reflect the energy-intensive nature of its industry, the relative inefficiency of its infrastructure, and the fact that it is big and cold. Another example is Bahrain, where energy consumption per head is 46 per cent higher than in the USA. But that is mainly because it refines (and therefore, according to the statistics, consumes) crude oil that it produces or imports, and has a very small population, not because its citizens consume more energy than the USA's do.

Electricity access is an important measure of inequality. Between 1970 and 2013 the number of people without electricity fell from 1.87 billion to 1.2 billion, while the global population rose from about 3.7 billion to about 6.8 billion. (See Table 3, p. 42.) Thus electricity was supplied to 2 billion extra people, in large part thanks to government-supported campaigns in China, India, Thailand, Brazil, Mexico and other countries. The level of electrification in the countryside outside the rich world – where it has historically been lowest – rose from 12 per cent in 1970 to 63 per cent in 2010. Still, in 2013, 17 per cent of the world population, including 68 per cent of sub-Saharan Africans and 19 per cent of Indians, had no electricity.[32] Those rural households outside the rich world that have electricity use very little compared to urban households, and many connections are precarious, either because households cannot pay for the supply and/or because the quality of networks is poor. (See pp. 109–10.) A different type of inequality is suffered by millions of people living in fuel poverty in rich countries. Many such people live in an environment where basic amenities, including heat, cooking fuel and electricity, are supplied on a commercial basis, but they are short of money to pay for it.[33]

In their overview of electricity supply outside the rich world, the *Global Energy Assessment* authors, taking into consideration erratic and poor-quality supply, concluded that there is an 'empirical dividing line between advanced and developing economies' at around 2000 kWh/year of electricity per person. 'The resulting "electrification gap" effectively excludes nearly *half* the world's population [their emphasis] from the potential benefits of a global economy.'[34]

The history of fossil fuel consumption is in part the history of these inequalities, and the reasons why they have persisted and in some respects increased.

4
Fossil fuel consumption in numbers

The growth of fossil fuel use since 1950 is set out in numbers in this chapter. First an overview of global consumption is provided, broken down by fuel and by parts of the world, and with respect to greenhouse gas emissions. Then consumption is outlined sector by sector. (Readers may treat the sector-by-sector descriptions as reference, looking at them briefly now and returning to them when reading Part Two.)

Most of the statistics used here are collected by the IEA, which has since the 1970s monitored the paths taken by energy supplies to final consumers. They provide a breakdown of consumption between different sectors (industry, agriculture, residential, and so on). Other international statistics say much less, if anything, about the sectoral breakdown. The World Bank's historical statistics[1] include those measuring energy consumption per capita and per unit of GDP. These allow comparisons between countries with respect to their energy use and level of economic activity, and potentially highlight energy efficiency issues, but provide little insight into the technological and economic systems that consume the energy.

Neither energy-use-per-capita, nor energy-use-per-unit-of-GDP statistics distinguish between direct and indirect consumption. Consequently, they do not reflect the effect of the export of energy-intensive industrial processes from the Global North to the South. Increasingly, products manufactured in the South are consumed in the North; the energy used in their production is consumed directly in the South, and included in the statistics there; its indirect consumption in the North is masked. To address this problem, researchers into the greenhouse gas effect of fossil fuel consumption devised methods of consumption-based accounting that attribute indirect consumption to the country where goods are consumed, rather than where they are made. So the energy used to produce a steel bar made in China but used in the USA is attributed to the USA, not to China.[2] This method has been useful to developing country governments in the negotiation of international agreements on greenhouse gas emissions. But again it provides only limited insight into systems that consume energy.

The consumption statistics cited here provide half the pieces of a jigsaw. Production statistics are the other half. In addition to such standard sources as the IEA and the *BP Statistical Review*, the Carbon Majors Database measures fuel production historically by companies, rather than countries. This shows the highly concentrated nature of fossil fuel production, attributing 71 per cent of fossil-fuel-generated carbon emissions since 1988 to just 100 companies.[3] (Methods of energy accounting are further discussed in Appendix 1.)

Global consumption: fuels and regions

An overview of global fossil fuel consumption since 1950 is provided in Table 4. Between 1950 and 1965, total consumption more than doubled, from around 1.6 billion toe/year to more than 3.5 billion toe/year. Between 1965 and 1970 total consumption grew by more than 30 per cent. Thereafter, consumption growth slowed: during the 1970s and 1980s it grew on average by 11.3 per cent in each five-year period. In the quarter century after 1990, it grew on average by 9.3 per cent in each five-year period, but the absolute volumes of extra consumption were larger than ever. The main interruptions in consumption growth were due to economic crises, such as the recessions of 1979–81 and the recession of 2008–09. In the early 1990s, the first international agreements on climate change were signed, but these made no impact on consumption growth. By 2000 total fossil fuel consumption was more than five times the 1950 level; by 2015, more than seven times.

The OECD countries, with less than one quarter of the world's population, consumed about three quarters of fossil fuels in 1950, and just under 70 per cent in 1965. (See Table 2, p. 19.) Thereafter, OECD countries' consumption grew less rapidly than others'. In 1990, the OECD, with little more than one fifth of the population, accounted for 55 per cent of fossil fuel consumption. Only in the late 2000s did the non-OECD share surpass 50 per cent. This did not, though, indicate a reduction in inequalities: it was mainly due to the relocation to developing countries of industries producing goods for rich countries.

The late twentieth century is often thought of as an age of oil that superseded an age of coal. Actually, coal use declined *relatively* while rising in *absolute* terms. Oil consumption overtook coal consumption in the 1960s, thanks largely to the expansion of rich countries' road transport. Oil was used in power stations, too, but not if coal could be mined nearby. To this day, coal is mainly used in the country where it is mined: in 2010, 14 per cent of it was traded across national borders, compared to about 60 per cent per cent of oil.[4] After 1970, oil consumption rose steadily in rich and poor countries alike, and the proportion of it used for road transport kept growing. But total coal use did not fall far behind. In 1970, the total oil consumed (counted by energy content) was about half as much again as coal; in 1990, about three-eighths greater; by 2015, only about one-eighth greater. In 2015, oil use was more than eight times its 1950 level; coal use was in second place, having grown four times over.

Gas was used in the post-war period mainly for household heating and cooking, and for industrial processes. It became a major electricity sector fuel

Table 4 Fossil fuel consumption, 1950–2015: overview

Millions of tonnes of oil equivalent

		1950	1960	1965	1970	1975	1980	1985	1990	1995	2000	2005	2010	2015
Total		**1567**	**2518**	**3539**	**4663**	**5362**	**6081**	**6376**	**7139**	**7448**	**8101**	**9334**	**10366**	**11306**
% rise on 5 yrs earlier		n/a	n/a	40.5%	31.7%	15.0%	13.4%	4.8%	12.0%	4.3%	8.1%	15.2%	10.0%	9.1%
Total	OECD	n/a	n/a	2457	3252	3484	3760	3629	3947	4214	4582	4780	4636	4494
Total	*non-OECD*	n/a	n/a	*1082*	*1410*	*1878*	*2321*	*2747*	*3191*	*3234*	*3519*	*4555*	*5731*	*6812*
Coal	OECD	956	1199	880	882	841	974	1081	1101	1059	1133	1178	1117	979
	non-OECD			*549*	*620*	*749*	*830*	*989*	*1110*	*1176*	*1210*	*1745*	*2347*	*2861*
Gas	OECD	151	368	449	673	739	821	803	905	1073	1226	1296	1406	1459
	non-OECD			*145*	*225*	*332*	*475*	*686*	*863*	*853*	*952*	*1203*	*1458*	*1676*
Oil	OECD	433	899	1128	1697	1905	1966	1745	1941	2082	2224	2305	2113	2056
	non-OECD			*389*	*565*	*796*	*1016*	*1071*	*1218*	*1206*	*1358*	*1606*	*1926*	*2275*

Source: Darmstadter, *Energy in the World Economy* (1950 and 1960), BP *Statistical Review* (from 1965)

Note: figures have been rounded to nearest mtoe.

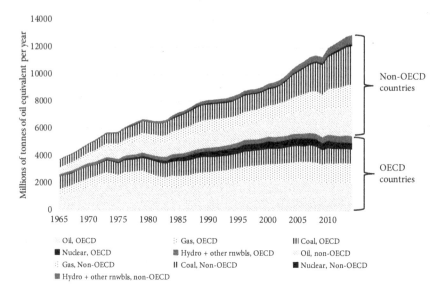

Figure 6 Consumption of commercially traded energy, 1965–2014
Millions of tonnes of oil equivalent per year

in the 1970s in the Soviet Union, and in the 1980s more widely. In the 1990s it often replaced coal in the electricity sector where there was direct competition. In the OECD countries, gas overtook coal as the second fossil fuel after oil in the early 1990s; in the non-OECD countries, by contrast, coal pushed oil into second place during the 2000s.

The shares of OECD and non-OECD countries in the consumption of all commercially-traded energy products – that is, including nuclear, hydro and other renewable energy used to produce electricity, as well as fossil fuels – are shown in Figure 6. Note the bulge in non-OECD coal consumption, mainly concentrated in China, in the 2000s, and the very temporary effect of the 2008–9 economic crisis, which dampened energy consumption for a year but then gave way to a new upward surge. Figure 6 is based on the *BP Statistical Review* and therefore does not include non-commercial biofuels. Figure 6 also shows the very modest role of renewable energy: in 2015, hydro (6.8 per cent) and all other renewables (2.8 per cent) together accounted for 9.6 per cent of primary energy consumption.[5]

Global warming and physical constraints

There are three causal links between fossil fuel consumption and the impacts of global warming. First, burning specific quantities of fossil fuels produces specific quantities of greenhouse gases; this relationship can be measured accurately with basic chemistry. Second, greenhouse gas emissions over time cause global warming. (Greenhouse gases in the atmosphere, including water vapour, carbon dioxide, methane and nitrous oxide absorb heat, prevent it from escaping into space, and emit it, including downwards, keeping the earth's surface warm and making it habitable. Adding to the volume of these gases makes this effect stronger, and makes the earth's surface warmer.[6]) Measuring the rate at which this happens is more complex, but several decades of paleoclimatology, computer modelling and other research have produced estimates that are not subject to controversy among climate researchers. Third is the causal relationship between warming, measured by, for example, the rise in global average temperature and a range of effects such as sea level rise, ice melt, weather volatility and changes in climate. This relationship is not in doubt, but the details are subject to greater uncertainties.

The figures used by the UN and other organisations – most obviously, the target of keeping global average temperature 'well below 2°C above pre-industrial levels' – reflect a political consensus about the required response to the global warming danger. The consensus is far from total. The Paris climate conference (2015), faced with scientific assessments that 2°C was more dangerous than previously thought, declared a lower temperature – 1.5°C higher than pre-industrial levels – to be a preferable target. But many governments regard 1.5°C as an absolute necessity, and many climate scientists urge aiming for a 1°C average temperature rise.[7] And of course these discussions do not reflect the attitudes of millions of people who are already directly affected by climate change. What is acceptable to politicians may not be acceptable to Bangladeshi fishing communities or tropical zone farmers, or millions of others whose opinions are not registered at international diplomatic gatherings. So these numbers are not presented here as inviolable norms.

In considering reducing fossil fuel consumption, a relevant analytical tool is the global carbon budget, an estimate of the amount of carbon that can be pumped into the atmosphere while keeping to certain temperature levels. The carbon budgets for the period since 1870, included in the most recent report of the IPCC, were: 2,900 billion tonnes of CO_2, to have a 66 per cent chance of limiting warming to 2°C; and 2,250 billion tonnes, to have a 66 per cent chance of limiting warming to 1.5°C. Most of these amounts have been used up between 1870 and the present. The remaining budget was estimated in 2011 at 1,000 billion tonnes, to limit warming to 2°C; or 400 billion tonnes, for 1.5°C.[8]

The amount of carbon dioxide (CO_2) emissions produced by burning fossil fuels depends on their quality. The dirtiest types of coal produce almost twice as much CO_2 per unit of energy as gas does: burning 1 toe of coal produces 4.1–4.2 tonnes of CO_2; 1 toe of oil products, 2.5–4.2 tonnes of CO_2; 1 toe of natural gas, 2.3 tonnes of CO_2, as measured by IPCC guidelines. The numbers vary with derivative products, non-fuel uses, and so on. Moreover, significant volumes of methane – the global warming effect of which is dozens of times greater than that of CO_2 over shorter time scales – are emitted during production and supply of natural gas (due to leakage, venting and flaring).[9]

In the mid-2010s, climate researchers found that fossil fuel use was producing about 36.3 billion tonnes of CO_2 per year, up from an average of 29.3 billion tonnes/year in 2000–2009. (A smaller amount, about one tenth of this volume, is produced by land use change; and these numbers do not include the effect of methane and other greenhouse gases.) In other words, the world economy is now producing, every three years, more than 100 billion tonnes of CO_2 – an amount that it took 15–20 years to produce during the post-war boom, and that is probably greater than the entire CO_2 emissions of the nineteenth century. Obviously, at the current rate, the economy would entirely use up the carbon budget considered safe by the IPCC, some time between the mid-2020s (for the estimate for 1.5°C) and the late 2030s (for 2°C).[10]

There is a consensus among climate researchers, international agencies and, increasingly, energy companies, that the carbon budget that can safely be used is substantially smaller not only than the amount of carbon contained in fossil fuel resources, but even than the amount contained in reserves deemed to be economically recoverable.[11] One climate research group found that the carbon budget (with a view to limiting global warming to 2°C) is equal to half of proven economically recoverable oil, coal and gas reserves; another put it at about 20 per cent of coal reserves, 50 per cent of gas and 70 per cent of oil; the Carbon Tracker Initiative put it at 20–30 per cent of indicated fossil fuel reserves.[12] In other words, society will run out of atmosphere into which it can pump greenhouse gases without potentially catastrophic climatic changes long before it runs out of fossil fuels; it will confront 'peak greenhouse gas emissions' long before reaching 'peak oil' (or gas or coal).

That does not mean resource availability is irrelevant. In recent decades, especially in the case of oil, costs have risen, as production has moved on from the 'low hanging fruit' (big fields with easy-to-produce oil) to so-called lower quality resources that require, for instance, deeper wells or additional processing. Technological change, such as the use of fracking, has counteracted and interrupted this trend, but not reversed it. Generally, each barrel of oil produced is more costly to access than the previous one. It also takes more energy to produce – and this tendency for the energy returned on energy invested (EROEI, usually shortened to EROI) to fall is a long-term constraint on fossil fuel use, albeit secondary to the urgency of avoiding dangerous climate change.[13]

Fuel consumption by sector

Oil, gas and coal are *primary energy* products. Some of them – coal, or gas for manufacturing fertiliser, or for cooking and heating – are used by companies or households without much processing as *final energy*. Others are transformed into secondary products that are forms of final energy: oil is refined into diesel, petrol, and so on; coal, gas and some oil products are burned to produce electricity and heat. (See Figure 3, p. 27.)

Tables 5-a and 5-b are adapted from IEA energy balances,[14] and provide snapshots of the world's energy consumption at 20-year intervals: in 1971, 1991 and 2011. Explanatory notes for the tables are marked ▶.

Table 5-a Global primary energy supply, and transformation of energy

World totals, millions of tonnes of oil equivalent	1971	1991	2011
Total primary energy supply (= sum of supply for transformation and supply for final consumption)	5527.8	8841.3	13128.9
Coal & peat	1442.0	2167.3	3785.3
Oil & oil products	2437.4	3252.2	4136.2
Gas	894.9	1724.8	2790.1
Nuclear	29.0	549.8	673.7
Hydro & other renewables	724.5	1147.0	1742.2
Other/statistical adjustment	0.1	0.1	1.5
Energy supply to final consumption (see Table 5-b)	3807.7	5160.3	7019.9
Energy supply for transformation	1720.1	3681.0	6109.0
Coal & peat	799.1	1423.4	2894.6
Oil & oil products	444.1	610.8	501.7
Gas	313.0	762.2	1421.7
Nuclear	29.0	549.8	673.7
Hydro & other renewables	134.9	334.8	617.3
Energy sources for transformation (= sum of energy used and outputs)	1720.1	3681.0	6109.0
Transformation: energy used			
Energy used in producing electricity	613.8	1353.9	2462.2
Energy used in combined heat and power plants	58.8	253.9	238.8
Energy used in producing heat	2.7	39.4	36.5
Gas works, artificial fuel plants & liquefaction plants	6.4	21.5	28.0
Energy used in petrochemical plants	0.2	0.6	0.5
Energy used in oil refineries	18.0	38.4	30.1
Other energy industry own use and losses	370.2	624.5	999.7
Blast furnaces	73.8	104.5	191.3
Coke ovens	95.2	39.7	58.7
Other transformation	22.9	42.5	74.0
Transfers and statistical differences	12.4	-33.4	123.4
Transformation: outputs			
Electricity	377.5	860.3	1587.3
Heat	68.3	335.6	280.0

Source: IEA Energy Balances

Table 5-b Global final consumption of energy

Millions of tonnes of oil equivalent		1971	1991	2011
Final consumption: energy sources		4253.5	6356.1	8885.7
Coal & peat		642.8	744.0	890.7
Oil & oil products		1993.3	2641.3	3634.5
Gas		581.9	962.6	1368.4
Hydro & other renewables		589.6	812.3	1124.9
Electricity		377.5	860.3	1587.3
Heat		68.3	335.6	280.0
Final consumption: uses of energy		4253.5	6356.1	8885.7
Industry	Iron & steel	202.7	231.9	452.4
	Chemical & petrochemical	147.1	255.0	352.9
	Non-ferrous metals	47.2	69.3	112.2
	Cement & other non-metallic minerals	108.2	165.1	332.4
	Machinery & transport equipment	86.4	104.6	167.8
	Mining and quarrying	14.9	36.0	74.3
	Food & tobacco	63.7	94.5	152.1
	Other industry	711.0	807.3	818.3
	Construction	22.9	32.1	46.3
Transport	Domestic aviation	56.2	91.5	97.8
	World aviation bunkers	56.5	84.2	159.8
	Road	602.6	1140.9	1844.8
	Rail	80.9	49.0	53.1
	Pipelines	17.5	59.5	64.1
	Domestic navigation & other transport	48.3	56.4	65.9
	World marine bunkers	108.4	119.2	207.1
Residential		1038.6	1559.5	2062.7
Commercial & public services		333.1	475.8	713.7
Agriculture, forestry and fishing		116.5	172.9	190.7
Other		163.0	267.2	129.2
Non-energy use	Chemical feedstocks	117.2	325.5	588.0
	Other non-energy use	110.4	158.8	200.0

Source: IEA Energy Balances

▶ The primary energy sources in the first section of Table 5-a, *Total primary energy supply*, are either used for transformation (to produce final energy, mainly electricity and heat), or go directly to final consumption, displayed in

Table 5-b. Nearly half of the primary energy supply goes for transformation, as you can see in Table 5-a, by comparing the rows *Primary energy supply to final consumption* and *Primary energy supply for transformation*.

▶ *Primary energy supply for transformation* is divided into two: (1) *Transformation: energy used*, which exits the energy balance during the transformation process (e.g. energy used when coal is burned to produce electricity); and (2) *Transformation: outputs*, as in electricity and heat that go directly to consumers. The sum of *Energy supply to final consumption*, and the *Transformation outputs*, is the *Final consumption: energy sources* in Table 5-b.

The largest item listed as *Transformation: energy used* is *Energy used in producing electricity*, which exits the energy system from power stations, mostly as waste heat into the atmosphere. The steam that rises from power stations' cooling towers is produced by this heat. (See p. 29.) The limits to power station and heat plant efficiencies are reflected in Table 5-a: the sum of the *energy used in producing electricity, energy used in combined heat and power plants* and *energy used in producing heat* has always been more than the *electricity* and *heat* produced. In other words, the aggregate efficiency of the world's electricity and heat production is less than half (40.55 per cent in 2011).

Electricity lost from transmission and distribution networks is the bulk of the row *Other energy industry own use and losses* in Table 5-a. The IEA has estimated that these losses account for between 7 per cent and 15 per cent of electricity supply, depending on local conditions; ABB, the engineering firm, has estimated them at 9 per cent. These losses are much greater outside the rich world, for example, 15 per cent in Latin America and 20 per cent in India. Between 1990 and 2011, the level of losses in rich countries fell thanks to technical improvements, while it rose elsewhere.[15] That row also includes some other energy used in transformation processes.

▶ For simplicity, Tables 5-a and 5-b include oil, and oil products, in single rows. Strictly speaking, oil products are secondary energy, produced by transformation. The row *Energy used in oil refineries* reflects the energy cost of that processing.

▶ *Transformation: energy used* also includes *Energy used in petrochemical plants*, and *Blast furnaces* and *Coke ovens* that produce heat (and some electricity) to make steel and other metals products. (Energy used further downstream in metallurgical processes is counted as final consumption.)

Energy use by final consumers is represented, albeit imperfectly, in Table 5-b. Among the energy sources are directly consumed fossil fuels; hydro and other renewables (most of which are the non-commercial biofuels consumed by people outside the commodified energy system); and electricity and heat. Electricity grew as a proportion of the total energy sources for final consumption from 8.9 per cent in 1971 to 17.9 per cent in 2011.

▶ In Table 5-b, *Final consumption: uses of energy*, includes *industry* (with the *other industry* row testifying to the poor quality of information); *transport*; and *residential* use. Most *non-energy use: chemical feedstocks* is gas, and some coal, used as raw material (rather than fuel) in chemical and petrochemical industries.

The energy balances of the USA, long the largest energy consuming country and now the second, are shown in Tables 6-a and 6-b; China's in Tables 7-a and

Table 6-a USA: total primary energy supply, and transformation of energy

Millions of tonnes of oil equivalent	1960	1971	1991	2011
Total primary energy supply (= sum of supply for transformation and supply for final consumption)	1019.3	1587.5	1930.6	2191.1
Coal & peat	222.9	279.4	455.0	479.1
Oil & oil products	467.1	722.0	741.0	786.0
Gas	283.7	516.9	458.5	568.6
Nuclear	0.1	10.6	169.2	214.1
Hydro & other renewables	45.1	58.3	105.0	140.2
Other/statistical adjustment	0.4	0.3	1.9	3.2
Energy supply to final consumption (see Table 6-b)	748.1	1105.1	1052.4	1157.5
Energy supply for transformation	271.6	482.7	880.1	1035.2
Coal & peat	130.4	200.9	400.1	454.5
Oil & oil products	35.7	88.7	71.9	40.4
Gas	92.2	159.0	158.1	257.0
Nuclear	0.1	10.6	169.2	214.1
Hydro & other renewables	12.7	23.2	78.9	67.7
Other/statistical adjustment	0.4	0.3	1.9	1.6
Energy sources for transformation (= sum of energy used and outputs)	271.6	482.7	880.1	1035.2
Transformation: energy used				
Energy used in producing electricity	109.9	219.0	418.9	506.7
Energy used in combined heat and power plants	0.0	0	49.1	23.3
Gas works, artificial fuel plants & liquefaction plants	0.4	0	1.1	0.8
Energy used in oil refineries	-4.5	-13.8	-2.9	-21.3
Other energy industry own use and losses	100.3	138.7	140.4	168.8
Blast furnaces	14.2	14.6	7.0	4.9
Coke ovens	5.5	3.6	3.4	2.3
Transformation: outputs				
Electricity	59.2	123.8	238.5	324.8
Heat	0	0	6.1	6.6

Source: IEA Energy Balances

7-b. There are interesting contrasts. The USA's high level of car use is reflected in the row *transport: road* (which includes freight, but is mostly fuel for private cars), which rose from 22.5 per cent of final energy use in 1960 to 34 per cent of a much larger total in 2011. China's road transport rose from 1.5 per cent of final energy use in 1971 to 11.4 per cent in 2011. The last column of Table 7-b (2011) provides insights into China's export-focused industrial boom. *Industry: iron and steel*, a mere 5.4 per cent of final energy consumption in 1991, leapt up

Table 6-b   USA: final consumption of energy

Millions of tonnes of oil equivalent		1960	1971	1991	2011
Final consumption: energy sources		806.9	1228.6	1295.1	1487.4
Coal & peat		92.5	78.5	54.9	24.6
Oil & oil products		431.4	633.3	669.1	745.7
Gas		191.5	357.9	300.4	311.6
Hydro & other renewables		32.3	35.1	26.2	72.5
Electricity		59.2	123.8	238.5	324.8
Heat		0.0	0.0	6.1	6.6
Statistical adjustment		0.0	0.0	0.0	1.6
Final consumption: uses of energy		806.9	1228.6	1295.1	1487.4
Industry	Iron & steel	45.3	40.5	17.8	17.7
	Chemical & petrochemical	7.4	18.9	67.6	55.4
	Non-ferrous metals	0.0	8.3	11.2	10.0
	Cement & other non-metallic minerals	0.0	6.3	10.1	19.6
	Machinery & transport equipment	0.0	9.2	14.7	29.8
	Mining & quarrying	0.0	1.8	2.8	4.7
	Food & tobacco	0.0	7.3	9.9	29.6
	Other industry	226.2	276.3	148.9	91.3
	Construction	0.0	0.0	0.8	6.4
Transport	Domestic aviation	23.0	45.7	61.1	49.3
	Road	181.8	293.4	386.5	507.1
	Rail	13.4	12.5	10.2	12.2
	Pipelines	0.0	17.3	14.1	15.9
	Domestic navigation & other transport	11.7	8.2	10.9	5.1
Residential		147.7	233.0	217.0	263.6
Commercial & public services		76.5	145.0	160.4	205.3
Agriculture, forestry and fishing		10.8	16.5	14.4	19.0
Other		25.5	17.6	22.1	11.5
Non-energy use	Chemical feedstocks	9.9	37.1	74.9	96.3
	Other non-energy use	27.7	33.6	39.6	37.9

Source: IEA Energy Balances

to 13.5 per cent of it in 2011. The rows *Blast furnaces* and *coke ovens* in Table 7-a must also be attributed to iron and steel production. The total 357 million toe was 12.9 per cent of total primary energy supply, not much less than the total commercial energy consumed by Chinese households.

Inequality in the world energy system is vividly reflected in Figure 7 (p. 66), derived from the IEA's energy balances for Nigeria. The columns show the sources of final energy used domestically on the left, compared (by energy content) to the crude oil used domestically and exported. In 1971, almost all the

Table 7-a China: total primary energy supply, and transformation of energy

Millions of tonnes of oil equivalent	1971	1991	2011
Total primary energy supply (= sum of supply for transformation and supply for final consumption)	394.6	856.8	2762.1
Coal & peat	192.0	506.7	1886.5
Oil & oil products	42.6	123.5	446.4
Gas	3.1	13.4	110.2
Nuclear	0	0	22.5
Hydro & other renewables	156.9	213.2	296.8
Other/statistical adjustment	0	0	-0.4
Energy supply to final consumption (see Table 7-b)	333.2	630.5	1242.8
Energy supply for transformation	61.3	226.3	1519.3
Coal & peat	51.6	183.6	1334.3
Oil & oil products	5.4	27.9	42.9
Gas	1.7	4.1	38.2
Nuclear	0	0	22.5
Hydro & other renewables	2.6	10.8	81.4
Energy sources for transformation (= sum of energy used and outputs)	61.3	226.3	1518.9
Transformation: energy used			
Energy used in producing electricity	24.9	116.0	622.1
Energy used in producing heat	0	3.8	20.6
Gas works, artificial fuel plants & liquefaction plants	0	1.1	3.5
Energy used in oil refineries	0.7	1.2	10.9
Other energy industry own use and losses	4.5	47.2	187.4
Blast furnaces	6.9	16.1	100.2
Coke ovens	4.5	7.0	35.2
Transfers and statistical differences	9.5	-28.0	138.0
Transformation: outputs			
Electricity	*10.3*	*47.3*	*335.8*
Heat	*0*	*14.6*	*65.1*

Source: IEA Energy Balances

energy used by Nigeria domestically was hydro and other renewables – which in Nigeria's case was more than 99 per cent biomass – while more than twice that total went abroad as crude oil exports. By 2011, little had changed. The oil used domestically, together with coal and peat, at least forms a visible item on the Figure; so (just) does the minute amount of electricity. The quantity of biomass used had meanwhile quadrupled. And the total of energy consumed in Nigeria is still exceeded by the amount shipped abroad in oil tankers. (The same information is presented in table form in Table 11, p. 209.)

Table 7-b China: final consumption of energy

Millions of tonnes of oil equivalent		1971	1991	2011
Final consumption: energy sources		343.5	692.4	1644.1
Coal & peat		140.4	323.1	552.2
Oil & oil products		37.1	95.6	403.4
Gas		1.4	9.4	72.1
Hydro & other renewables		154.3	202.5	215.4
Electricity		10.3	47.3	335.8
Heat		0	14.6	65.1
Final consumption: uses of energy		343.5	692.4	1644.1
Industry	Iron & steel	4.7	37.2	221.6
	Chemical & petrochemical	2.1	42.3	112.6
	Non-ferrous metals	0	8.3	46.7
	Cement & other non-metallic minerals	0.4	58.8	165.2
	Machinery & transport equipment	0.9	26.6	66.3
	Mining & quarrying	0.7	6.0	17.1
	Food & tobacco	0.3	18.2	28.2
	Other industry	103.6	54.2	112.7
	Construction	0.2	5.7	15.0
Transport	Domestic aviation	0	1.0	11.1
	Road	5.1	25.1	186.9
	Rail	6.5	10.1	12.2
	Pipelines	0	0	0.1
	Domestic navigation & other transport	3.2	3.5	25.6
Residential		194.3	295.8	364.3
Commercial and public services		4.3	14.1	56.1
Agriculture, forestry & fishing		11.5	30.2	33.5
Other		4.6	14.6	57.5
Non-energy use	Chemical feedstocks	0	19.7	58.2
	Other non-energy use	1.1	20.8	53.0

Source: IEA Energy Balances

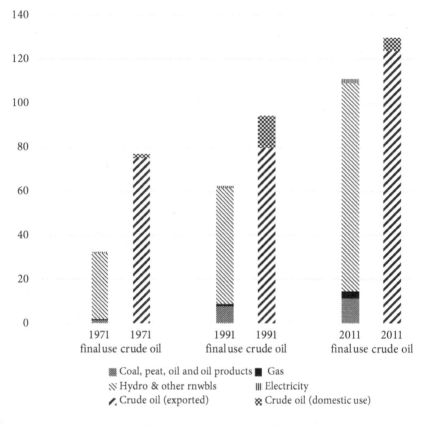

Figure 7 Nigeria: domestic final use of energy, and crude oil supply
Millions of tonnes of oil equivalent

Note: Domestic use of coal, peat, oil and oil products have been included as one item so that it is large enough to be visible on the chart. More than 99% of the 'hydro and other renewables' comprises biofuels and waste, i.e. fuel wood and other non-commercial fuels.

Source: IEA energy balances

Electricity and heat

The large, and growing proportion of fossil fuels used to produce electricity and heat is shown in Table 5-a. The amount of electricity produced globally more than quadrupled between 1971 and 2011, thanks to the advance of electrification on one hand, and more intensive use of electricity in rich countries on the other. (See the row, *Electricity*, in Table 5-a.) The OECD's share of global electricity consumption fell, from 65 per cent in 1985 to 45 per cent in 2015, but the rich countries remained far more electricity-intensive than the rest. In 2005 the OECD used 8,053 kwh per person per year, almost ten times the 83 kwh per person per year used in the least developed countries.[16]

▶ The energy consumed by the world's power stations and combined heat and power plants is in the rows *Energy used in producing electricity* and *Energy used in combined heat and power plants*. They produce *Electricity* and *Heat*.

By far the largest user of electricity globally is industry, whose consumption rose from 204.3 billion toe in 1971 to 673.4 billion toe in 2011. The largest industrial electricity users were and are also among the largest direct consumers of fossil fuels: iron and steel making (28.6 billion toe in 1971, rising to 91.4 billion toe in 2011) and the chemical and petrochemical industries. Aluminium production also uses large quantities of electricity. Industry's share of electricity consumption fell between 1971 and 2011, from 54.1 per cent to 42.4 per cent, mainly because other sectors' consumption rose even faster than industry's. The most significant were residential use, which rose from 85.7 billion toe (22.7 per cent) in 1971 to 428.6 billion toe (27 per cent) in 2011, and commercial and public services, an IEA category that embraces both the world's small businesses, and public organisations and buildings (schools, hospitals, and so on), which rose from 55.1 billion toe (14.6 per cent) to 363.3 billion toe (22.9 per cent) in the same 40-year period. In the 2000s, technologies of the third Industrial Revolution produced a giant new source of electricity demand. (See p. 162.) World electricity consumption by sector is summarised in Table 12. (See p. 210.)

The second intermediate form of energy produced by the transformation sector is heat, usually supplied to industry or to homes via hot-water-based urban heating systems. Its share of final energy consumption doubled from 1.6 per cent in 1971 to 3.1 per cent in 2011.

▶ These amounts are in the row *heat* in *Final consumption: energy sources*, in Table 5-b.

Industry

The expansion of fossil fuel use by industry since 1950 falls into two periods: up to the mid-1970s, and afterwards. During the post-war boom, it leapt up in the rich countries. (See p. 82.) From the 1970s onwards, much energy-intensive industry moved outside the rich world. Energy demand for manufacturing, in OECD countries, fell by 15 per cent between 1973 and 2000; it continued to grow elsewhere. As a proportion of national fossil fuel use, consumption by industry fell in many rich countries, and grew elsewhere: in the USA, the share fell from 30 per cent in 1971 to 18 per cent in 2011, while in China it rose from 33 per cent to 48 per cent.[17] Five energy-intensive materials – steel, cement, plastics, paper and aluminium – dominate fossil fuel use by industry, and by the

2000s accounted for about half of it. By the 2000s, more than one third of the world's steel, and most of the cement and aluminium were being made outside the OECD.[18] (See p. 158.)

Among industrial sectors, iron and steelmaking is the largest fossil fuel user. Global crude steel output, which had been 28 million tonnes in 1900, rose to 192 mt in 1950, 770 mt in 1990 and 1,670 mt in 2014.[19] Including the coal (or, sometimes, gas) consumed by blast furnaces and coke ovens, steel production swallowed 302 million toe in 1971, rising to 702 million toe in 2011. After steel, the aluminium industry is one of the largest fossil fuel users, even though aluminium smelters often rely on hydro or nuclear power. Globally, production rose 30 times over between 1950 and 2007.

Since 1950, the biggest use of steel by far, worldwide, has been for construction. Construction, vehicles and industrial equipment were the dominant sources of steel demand in rich countries during the post-war boom. The trend persisted in 2012, when 42 per cent of steel went into buildings, and another 14 per cent into other infrastructure (e.g. roads and bridges). Other big uses were manufacture of vehicles (12 per cent) and electrical and mechanical equipment (16 per cent).[20]

Since 1950, overall, steel demand has fallen, relatively, in rich countries, as the role of heavy industry and construction in their economies shrank; outside the rich world, it has risen. Steelmaking has become more energy efficient: the world average amount of energy needed to produce a tonne of steel fell from around 1.2 toe/tonne in 1975 to around half that in the 1990s. In the mid-twentieth century, engineers' efforts were focused on reducing the amount of coking coal needed by blast furnaces. But energy is needed later in the process, too, e.g. to reheat steel to be rolled into semi-fabricated products.[21] The adoption from the 1980s of electric-arc furnaces, and technologies such as continuous casting, has reduced losses at these later stages.

Cement making, like steel making, was concentrated in the rich countries up to the 1970s. From the 1980s, rapidly expanding consumption in developing countries overtook the rich countries, with China leading the way. World cement consumption rose from about 10 million tonnes/year in 1910 to 600 million tonnes (mt)/year in 1970 and 2,800 mt/year in 2010.[22]

▶ Cement manufacture accounts for most of the row *Cement and other non-metallic minerals*, in Table 5-b, which rose from 2.2 per cent of total energy use in 1971 to 3.1 per cent in 2011.

Chemicals and petrochemicals production is the second-largest industrial consumer of fossil fuels after metallurgy. From the 1980s, cross-border trade swelled, but – in contrast to much materials production, which moved out of the rich world – this capital-intensive and increasingly high-tech sector largely stayed put and exported finished products. Even in 2005, the EU, the USA, Japan and China accounted for three-quarters of chemical production, and the EU, the USA and Japan hosted 28 of the headquarters of the 30 largest producers. Then a shift to the Gulf states and east Asia gathered pace, and Saudi, Qatari and Taiwanese companies joined the top rank. Most of the energy in chemicals production is used to make intermediate materials, such as ammonia (mostly to produce fertiliser, see below), chlorine and alkalines (used in industrial processes), and ethylene (for plastics). Other final products include pharmaceuticals, refrigerants, paints, solvents, soaps and synthetic fibres.[23]

▶ The chemical and petrochemical industries use fossil fuels as raw material, as well as to produce process heat and drive machinery. In Table 5-b, the fuel is in the row *Industry – chemical and petrochemical*, and the raw material in *Non-energy use – chemical feedstocks*.

Plastics are chemically fabricated, almost entirely from gas, oil products or coal. The big three plastics, polyethylene, polypropylene and polyvinyl chloride, together comprise 60 per cent of world output. The availability of cheap fossil fuels drove the industry's expansion: global production was just 50,000 tonnes/year in the 1920s; it reached 1 million tonnes (mt) in 1949 and 6 mt in 1960, and then leapt up to 100 mt in 1989 and 311 mt in 2014.[24] The biggest uses of plastics are for packaging and construction: in Europe, an industry association found that these accounted for 40 per cent and 20 per cent of output respectively. A global survey showed that packaging was the largest use, accounting for 26 per cent, but that little accurate information was available. Plastics consumption in the 2000s reached about 120 kg/person per year in rich countries; much of this is for packaging in the supply chain before goods reach the consumer. Of the 50 kg/person of plastic packaging produced each year in the UK, 30 kg is used moving goods from factories to shops. Among personal uses, plastic bottles accounted for 7 kg/person, and supermarket plastic bags for 1 kg/person.[25] Plastics, very few of which are recycled, contribute to the global waste crisis. (See pp. 75–6.)

In addition to energy-intensive materials production, fossil fuel consumption is also used in manufacturing industry for process heat, to drive machinery, and to produce electricity for lighting, machinery, and electronics and computers.

▶ In Table 5-b, *Other industry* reflects the poor state of statistics collection. Some national statistical agencies do not work out sufficiently detailed breakdowns by industrial sub-sector; where the sub-sector is unknown, the fuels may be included in this row.

The military

Fossil fuel consumption by the military grew substantially in the second half of the twentieth century, as warfare became more energy intensive. New generations of military aircraft and vehicles gulped down jet fuel and petrol at unprecedented rates; bigger-than-ever logistics infrastructures supported modern armies and air forces; and weapons manufacture used significant amounts of fuel.

It is almost impossible to estimate the level of military fossil fuel use. The USA, the world's largest military power, publishes limited information on its forces fuel procurement; many other countries publish no information at all. Researchers have estimated that, during the cold war (1950s–1980s), military purposes (including weapons manufacture and nuclear bases) accounted for at least 5 per cent of primary energy consumption in the USA and Soviet Union; that in the USA that level rose to 15–20 per cent during wars; that in the 1980s, military fuel use (excluding weapons manufacture) accounted for 2–5 per cent of primary energy consumption in the USA and the Netherlands, and 1.44 per cent in West Germany; and that, worldwide, nearly one quarter of jet fuel is used for military purposes.[26]

Armed forces have completely changed the way they use fuel. The energy specialist Sohbet Karbuz has estimated that the US military consumed, per serviceman or woman per day, 3.8 litres of petrol in the Second World War; 34 litres in the Vietnam war; 38 litres in the 1991 Gulf War; and 57 litres in 2007. In 2006, the US Air Force (then active in Iraq and Afghanistan) burned 9.85 billion litres of jet fuel – the same amount as all US aeroplanes consumed during the whole of the Second World War. The ability to pour ever-greater quantities of fuel into vehicles has been a factor in US military dominance. A US armoured division of 348 tanks uses 2.3 million litres of fuel per day, and a carrier battle group 1.6 million litres; a F-16 fighter jet burns as much fuel in an hour as a motorist might use in two years.[27]

The US Department of Defense was in the 2000s the world's biggest single consumer of commercial energy, and a larger consumer than Nigeria. Most of this is oil products: in the 20 years up to 2006, its reported annual consumption of those fell from 21.5 million toe/year to 13.4 million toe/year – but these figures only show part of the picture. They do not include weapons manufacture, military activities by the Department of Energy, NASA space agency or military contractors, or fuel used by US forces overseas.[28]

▶ Table 5-b does not accurately reflect military consumption. Some is included in the row *other*; some may be in *other transport* and *other industry*. (The IEA asks national statistical agencies to report energy consumption by the military under 'non-specific', which is included here in *other*, but many do not do so.)[29]

Transport

Energy use by road transport (overwhelmingly oil products, mostly in private cars) more than tripled between 1971 and 2011, and rose as a proportion of total final energy consumption from 14.2 per cent to 20.8 per cent. Aviation's share also rose, from 2.6 per cent to 2.9 per cent, while the share for rail, largely pushed out of the freight sector by trucks, fell from 1.9 per cent to 0.6 per cent.

The main driver of this fuel consumption growth was the rising car population. The world had about 50 million motor vehicles in 1950, 400 million in 1980 and 780 million in 2000. The rich countries were dominant, and even in the 2000s the developing countries were of secondary importance: non-OECD countries' share of the world's car fleet grew from a negligible level in 1960 to about 30 per cent in 2000. The gulf in car ownership per person remained: in 2000, there were 621.9 cars per 1000 people in North America, 140.6 in the former Soviet Union (after 10 years of especially rapid motorisation), 27.7 in Africa and 15.4 in South Asia.[30] Automobile dependence and the construction of car-based cities have assured the continued growth of fuel use for transport. The resistance by rich-country political elites to reducing subsidies to car drivers or to regulating fuel efficiency and vehicle sizes and weights, has played its part. (See pp. 83–4, 102, 127–8 and p. 165.)

After passenger cars, freight trucks are the second largest user of fossil fuels for transport: they consumed just under a quarter of the transport sector's total in the 2000s. Railroads have always been much more fuel-efficient than trucks, and have become more so over time. In the 1970s in the USA, trains on average used 1.4 toe to move one tonne of freight one kilometre, while boats used 1 toe and trucks used 4.8–15.3 toe; 40 years later in the 2000s, in Japan, railroads used just 0.5 toe/tonne/km, while trucks averaged 8.8 toe/tonne/km. Nevertheless, as cross-border trade ballooned from the 1980s, it was road freight that expanded at railroads' expense.[31]

Fossil fuel consumption by aeroplanes has risen steadily since the mid-twentieth century, boosted by the international agreement not to tax aviation fuel. (See p. 23.) Between 1960 and the mid-2000s, passenger air traffic grew at about 9 per cent per year, about 2.4 times faster than the world economy. The annual distance flown by passengers on scheduled flights rose from 40 billion km in the early 1950s to nearly 3 trillion km in 2000 and more than 6 trillion km in 2014. Since global warming was discovered, aviation has expanded even more rapidly than before. Between 1985 and 2005, the number of international flights quadrupled. Between 1990 and 2010, CO_2 emissions from aviation and shipping grew by around 80 per cent, around double the rate for the economy as a whole.[32]

▶ In Table 5-b, the rows *world aviation bunkers* and *world marine bunkers* are fuels supplied to aeroplanes and ships that make international journeys.

Agriculture

There are four significant areas of fossil fuel consumption in agriculture: (i) fertiliser manufacture; (ii) transportation of food; (iii) processing and packaging; and (iv) supply of electricity, heat and mechanical energy to industrial agriculture.

▶ In Table 5-b, (i) comprises part of *industry: chemical & petrochemical* and *non-energy use: chemical feedstocks*. *Agriculture, forestry and fishing* in Table 5-b probably comprises part, but not all of (iv). There is no way of disaggregating (ii) and (iii) from other uses.

Fertiliser manufacture. The expansion of industrial agriculture, most significantly in the USA during the post-war boom, drove up world production of ammonia, the key intermediate product for making synthetic nitrogen-based fertilisers, mostly from natural gas. World ammonia production grew from under 5 mt/year in 1950, to 31 mt/year in 1961 and 178 mt/year in 2010. In the 1970s, the Soviet Union developed its own chemical fertiliser production complex; fertiliser consumption soared in Asia and Latin America, to supply large-scale farms producing high-yielding types of rice, wheat and corn. Techniques developed in the USA in the 1950s went global, such as siting plants near gas wells and pipelines and integrating ammonia synthesis with other types of fertiliser production.[33]

Food transportation has been estimated to account for up to one quarter of all transportation, with the proportion even higher outside the rich world. With the widespread commodification of food, and the growth of food trade, *processing and packaging* also make a big contribution.[34]

Fossil fuel use is integral to *industrial agriculture*, and in particular livestock farming, and the energy-intensive production of high-protein feed for livestock. A researcher for the UN Food and Agriculture Organisation explained:

> Fossil energy is used for the production of feeds (land preparation, fertilisers, pesticides, harvesting, drying etc), their bulk transport (rail and/or sea freight), storage (ventilation) and processing (milling, mixing, extrusion, pelleting, etc) and their distribution to individual farms. Once on the farm, and depending on location (climate) season of the year and building facilities, more fossil energy is needed for the movement of feeds from the storage to the animal pens; for control of the thermal environment (cooling, heating or ventilation); and for animal waste collection and treatment (solid separation, aerobic fermentation; drying; land applications etc). Transport of products (meat animals to abattoirs; milk to processing plants; eggs to storage), processing (slaughtering, pasteurisation, manufacture of dairy products),

storage and refrigerated transport also require fossil fuels. Finally the distribution to the consumer and the final cooking process may also require expenditures of fossil fuels.[35]

These complexities have challenged researchers who have tried to work out how much fossil fuel energy is embedded in industrial agriculture outputs. On average, 25 units of fossil energy inputs go into animal protein production for one unit of energy output – eleven times more than for grain protein production, two specialists concluded in 2003. They found 4 kilocalories of fossil energy was embedded in each kilocalorie of broiler protein in a broiler chicken; for turkey it is also 4:1; milk and pork 14:1; eggs 39:1; beef 40:1 and lamb 57:1. In the case of beef and lamb, if the animals were fed only on good quality forage (and not on grain), the energy inputs could be reduced by half.[36]

Food consumption is unequal, and uses fossil fuel inputs unequally. Consumption of industrially produced meat in rich countries is especially fossil-fuel-intensive – and, where excessive, contributes to an obesity epidemic affecting 600 million people, while more than 2 billion others are undernourished. In recent decades, meat fed to rich-world domestic pets has become statistically significant too. Moreover, between a quarter and a third of all food produced is wasted, primarily in households, shops and restaurants in rich countries, and in poor-quality warehouses and transportation in poor countries.[37] There is scope for reducing fossil fuel use not only in food consumption and distribution, but also production, according to agronomists – by changing agricultural practices, reducing the level of fertiliser inputs, and so on.[38]

Fossil fuel consumption for agriculture may be between 5 per cent and 10 per cent of global final consumption of energy, but the IEA statistics do not allow a more accurate estimate.[39] (In addition, there are substantial greenhouse gas emissions from agriculture not related to fossil fuel use.[40]) Moreover, non-fossil energy sources (biofuels) for agriculture are probably hidden in the row *Residential*.

Buildings

The construction of buildings is a big energy consumer – although, even today, in developing countries many buildings use negligible amounts of energy-intensive materials. Industrialised construction methods in rich-country cities are the most energy intensive. Reinforcing bar (rebar) for concrete and masonry has accounted for a growing share of steel demand. Over-specification and use of generic designs, reflecting relative costs of material and labour, has pushed up demand for energy-intensive steel products.[41]

Once buildings are up, space heating has historically been the largest user of energy. During the post-war boom, it accounted for most (60–80 per cent) of energy use in buildings in rich, colder countries. Air conditioning, long insignificant outside the USA, has grown in the last three decades. But the share of both heating and cooling has fallen, as that of other electrical appliances has grown. In the USA, heating, air conditioning and ventilation accounted for a little under 70 per cent of energy use in homes in 1950–78, but only 36 per cent of energy use in buildings in 2006. The rich-world energy price shocks of the 1970s stimulated marked improvements in insulation standards in the 1980s – but even after these, architects and designers remained vocal about enormous *additional* potential efficiencies in heating buildings, and by the 2000s were pointing to potential energy savings of 90 per cent with existing technologies.[42]

Households

Energy consumption by households reflects the gulf between rich and poor. Even today, biofuels are *the leading* household energy carrier worldwide. Outside the rich world they account for seven-tenths of household energy consumption. These non-fossil fuels (wood, animal dung and other biomass) are consumed largely outside the commodified energy system, mostly for cooking. In second place is electricity, which has comprised a steadily growing share of rich-world household energy consumption. The next largest household energy carriers are gas and petroleum products.[43]

The rich-poor divide is also reflected in the size of households, which in turn conditions the ways that energy is consumed. In 1975, in OECD countries there were on average three people per household, and elsewhere, 5.6 people. Since then, the number of people per rich-world household has fallen further, due to changes in family structures, while the amount of space to be heated in each household has grown. In the Netherlands the average number of people per household fell from just under four in 1950 to a little over two in 1997; in the UK, the average size of homes rose from 140 square metres in the 1970s to 215 square metres in 2003.[44]

There are four broad categories of energy services used in households: 1. Heat for cooking. 2. Lighting, space heating and hot water. 3. Energy for electrical and other appliances that substitute for, or change the character of, domestic labour, such as refrigerators, washing machines and dishwashers. 4. Energy for appliances for communication or leisure (e.g. TVs, phones and computers). The third and fourth categories overlap. *Heat for cooking* is the most basic use of energy, even in the poorest rural households outside the rich world. Poor urban households often move from biofuels to such fossil fuels as liquefied petroleum gas (LPG), coal and kerosene. Industrial construction of homes brings natural gas and electricity. *Lighting* was the first energy service provided when homes were electrified, from the late nineteenth century. *Space heating and hot water* became common in the rich world only in the 1960s and 1970s, and outside the rich world remain today the comforts of small urban minorities.

The *energy used for electrical and other appliances* has since 1950 transformed the labour process in the home no less than in the workplace. In rich and poor countries alike, once households gained access to electricity and gas, they first bought appliances that eased the most physically demanding tasks: cookers, refrigerators and washing machines. (See pp. 90–2.) In rich countries, the main driver of higher household energy consumption since 1950 has been the proliferation of appliances. In Switzerland, for example, per capita electricity use tripled between 1960 and 1997, because the diffusion of appliances and the smaller number of people per household cancelled out efficiency improvement savings many times over.[45]

▶ In Table 5-b, household use is in the row *Residential* (23–5 per cent of total final energy consumption in 1971–2011). Workplaces, offices, and so on are in *Commercial and public services* (7.5–8 per cent).

Waste

Mass manufacturing techniques, urbanisation, and consumerism have stimulated the expansion of waste disposal into a major industry; significant quantities of fossil fuels are embedded in this waste. Most waste is produced in the rich world, mostly by industry; post-consumer waste (that is thrown away by individual consumers) has been estimated at one-fifth of the total. From the 1980s, the export of waste, including toxic waste from the rich world to developing nations for processing, or dumping, grew rapidly, and triggered attempts at regulation in the 1990s. In the 2000s, about 1.3 billion tonnes/year of municipal waste was being produced, with the amount per head from OECD countries running at twice the level of most developing countries and nearly four times Africa's.[46]

Wasteful packaging is a relatively recent rich-country phenomenon. In the UK, refillable (mostly glass) milk bottles reigned supreme (with 94 per cent of the market) in 1974, but had slumped to less than 10 per cent in 2006, being replaced by throwaway containers. In 1961, one third of the UK's beer was delivered in refillable containers; in 2006, less than 1 per cent. Refillables crashed out of the soft drinks market in the 1980s, falling from 45 per cent in 1980 to 10 per cent in 1989.[47] From the 1980s, plastics packaging became standard. Hardly any of it was recycled – a trend that could have been reversed by regulation. In 2013, 40 years after the launch of the repackaging symbol, only 14 per cent of plastic packaging was recycled; 40 per cent was going to landfill; and 32 per cent was 'leaking' from the collection system into the ground or the sea. The damage to oceans is palpable: plastics comprise nearly two-thirds of all the waste collected in coastal clean-up operations, one report showed; 5.25 trillion pieces of plastic, weighing more than 260,000 tonnes, accounts for 70 per cent of all garbage floating in the oceans, said another; and this floating rubbish is dwarfed 10–30 times over by the mountains of plastic refuse on the ocean floor, a third pointed out.[48]

In Part Two, the history of fossil fuel consumption since 1950 is surveyed chronologically.

Part II

Chronologies

5
The 1950s and 1960s: Post-war boom

Between 1945, when the Second World War ended, and 1973, the large capitalist economies expanded in a way not seen before or since. The world economy more than tripled in size. The engine room of this boom was the USA, which in 1950 accounted alone for between a quarter and a third of the gross world product.[1] The dollar became the reserve currency in the financial system established at Bretton Woods in 1944; and the international financial institutions, based in Washington, were strongly influenced by US policy. The USA also dominated the geopolitical order and maintained the 'balance of power' with the Soviet Union. In terms of industrial development, European, Japanese and Soviet industry raced ahead to compete with the USA; in terms of living standards, the gulf between rich and poor persisted.

Cheap fossil fuels were key building blocks of the boom. They supported the mushrooming mass production of energy-intensive raw materials; they poured into industrial agriculture; they underpinned new trade networks and fuelled fleets of private motor cars. The types of consumption established in the rich countries during the boom were reproduced across the world subsequently. In the 1950s, political and corporate elites assumed that fossil-fuel-based energy systems, supplemented by nuclear power, would and should expand at an accelerated rate for the foreseeable future. The 'dramatic advance of the human race over the past two hundred years' could be reversed if fossil and nuclear fuel supplies were not made ready, a key report to a UN gathering in 1955 declared. Oil would have to fill the gap while nuclear was developed, a senior Shell manager told the Fifth World Power conference the following year.[2]

The increases in fossil fuel consumption levels during the boom were manifestations of the 'great acceleration' of human impacts on nature. Oil led the way. World output of oil and oil products grew nearly fourfold between 1950 and 1968; natural gas output grew more than fourfold. By 1968, oil and gas accounted for more than 60 per cent of commercial energy consumption. Coal production had also grown by 45 per cent, but oil overtook coal as the most widely used fuel. (See Table 4, p. 55.) Oil and US power were closely bound together. Of the seven international oil companies that dominated production and distribution, the 'seven sisters', five were based in the USA.[3]

In this chapter, the supply of fuels in the boom is surveyed, and then consumption in world regions and sectors. There are comments, in a concluding section, on household fuel consumption and its effect on domestic labour.

Fuel supply: the rise of oil

During the boom, cheap oil became central to new ways of living and working in the rich countries. The spread of cars meant workers could live further from their workplaces and enjoy new types of weekend leisure. Long-distance freight became so cheap that rich-country populations could eat out-of-season fruit from other climatic zones. Oil was more capital-intensive than coal but less labour-intensive. Oil prices were low – and fell, overall, during the post-war boom. Oil products often replaced coal in power stations and domestic heating devices.[4] The rise of oil was also driven by suppliers. The large corporations had a near-monopoly on production, processing and distribution, and could manage pricing to their advantage. Economies of scale helped them; the USA and other states protected and supported them.

This relationship between the state and big oil was a central feature of the period.[5] Before and during the Second World War, the USA had been the world's largest oil producer. Its oil companies were supported by a wealth of tax breaks and subsidies. (Tax benefits to US oil and gas producers amounted to 13 per cent of their sales prices, compared to 1 per cent for hydro and solar energy.[6]) During the post-war boom, the geography of oil production shifted. Output grew rapidly in the Middle East (especially in Iran and Saudi Arabia) and North Africa (especially Libya). The Middle East produced less than one-fifth of the world's oil in 1950, and more than one-third by the early 1970s. (See Table 13, p. 210.) The USA and European powers sought political and economic control of the region, to secure oil supplies and to pursue Cold War aims against the Soviet Union, which was also becoming a major oil producer.

In 1951 the Iranian government, headed by Mohammed Mossadegh, nationalised the oil industry; the USA and Britain responded by supporting a coup that removed him in 1953. Production and export were put in western companies' hands. In Saudi Arabia, economic 'soft power' was used. In response to Saudi government demands for a greater share of oil sales revenue, the USA brokered a deal under which its oil companies could treat royalties paid to foreign governments as a tax-deductible expense. This foreign tax credit was, during the 1960s, larger than the oil companies' total tax bill.[7] The western powers were always ready to resort to military means to protect oil supplies. This fuelled their conflict with nationalism in the Middle East in general and the 1956 Suez crisis in particular.

In the USA, oil consumption had already overtaken coal by 1950; natural gas, for which the USA and the Soviet Union developed distribution systems in the post-war period, was also significant. In western Europe and Japan, oil overtook coal between 1960 and the mid-1970s. As a proportion of fossil fuels consumed, in 1950 hydrocarbons (oil and gas) already comprised 57 per cent in the USA but only 11 per cent in Europe; by 1968 hydrocarbons comprised 77 per cent in the USA and 61 per cent in Europe.[8] Cheap oil became cheaper, for one thing because the cost of transporting it fell as the size of tankers grew. By 1960, newly-built tankers were on average more than three times, and in some cases more than ten times, larger than those at sea in the mid-1950s. The European refining industry grew, so that crude oil from the Middle East could be processed in Europe.[9]

Support for electrification was another key function of the state. Since electricity was considered a strategic asset, the post-war settlement precluded national ownership of it in Germany and Japan, the defeated nations; in the USA, private corporations had controlled electricity generation and distribution from the start. But these were exceptions. Elsewhere, electrification was seen as a key to economic development and the state took the initiative. In 1946 the UK nationalised electricity generation and distribution and France nationalised generation. The state took part-ownership of electricity networks in Italy, Ireland, Greece and Portugal. Among developing countries, Mexico took majority control of the main electricity company (1946); so did India (1947), Egypt (1947), South Africa (1948) and Brazil (1959–63). From 1948, the World Bank made loans for grid development, usually by state-owned firms. Nationalisation was associated with the technological centralisation of previously disparate systems. Moreover, centralised coal-fired power stations became much bigger, and much more efficient. The capacity of individual turbine-generator sets leapt from 30 MW to 300 MW in less than a decade. New materials and fabrication methods for turbine blades, bearings and other components boosted performance and cut capital costs.[10]

The consumption of commercial energy during the post-war boom was heavily concentrated in the USA, Canada, western Europe, Japan and the USSR. In 1950, these countries, with 27 per cent of the world's population, accounted for 85 per cent of commercial energy consumption. By 1968 these shares had fallen slightly to 25.7 per cent and 78 per cent.[11] The USA and Canada developed energy-intensive economies that were in a league of their own, even in comparison to other rich countries. In 1950, North America's level of commercial energy consumption per capita was 3.7 times higher than Europe's, and 4.2 times higher than the Soviet Union's. Even in 1968, after two decades of European economic catching-up, the USA's level was 2.9 times higher than France's and 1.8 times higher than the UK's.[12]

Energy consumption relative to economic output was 'pervasively higher' in the USA and Canada than in other industrialised countries, a comparative research project found. The main causes were (i) structural differences (e.g.

the way industry, transport and buildings consumed energy); and (ii) US and Canadian government policies that kept post-tax energy prices far below those elsewhere. So in 1970, consumers in the UK paid 1.8 times as much, in France and Germany 1.7 times as much, and in Japan more than twice as much as those in the USA. Between 1960 and 1976, the energy efficiency of electricity generation fell in North America while it improved elsewhere.[13]

Industry and agriculture

Industry accounted for around three-eighths of world fossil fuel consumption during the boom years, a larger share than either transport or residential consumption.[14] Among industrial sectors, steel and cement manufacture were the most significant. Other rich countries raced to catch up with the USA. Japan, Italy and the Netherlands registered spectacular growth in energy consumption for industry (tenfold, sevenfold and sixfold respectively between 1950 and 1977). In the USA and western European countries, the fastest growth was already in the past; consumption roughly doubled in the USA, France and Belgium over the same period, and grew still more slowly in the UK.[15]

Since before the Second World War, and up to the early 1970s, fossil energy was cheap relative to labour in the rich countries. This stimulated the proliferation of energy-intensive manufacturing techniques, which in turn drove up levels of consumption. The electrification of rich-world industry played a part in this process. A survey of US energy prices showed that workers' wages became steadily costlier to businesses, by comparison with electricity, between 1935 and 1970; 'during the gradual shift to reduce total labour costs, machines and energy were introduced as substitutes for labour'. From 1950, electricity – rather than labour – tended to be substituted for capital, where it could be. One of the first detailed studies of post-war energy consumption, published in 1982, noted that 'low oil prices probably encouraged, until 1973, the development of energy intensive techniques and of low [energy] efficiency installations', especially in light industry.[16]

Demand for steel was driven by carmakers, industrial equipment manufacturers and construction firms. Cars gulp down energy when being made, as well as on the road, and between 1946 and 1973, car production grew almost tenfold. In 1946, the USA produced 80 per cent of cars; by 1973, the USA and western Europe each produced about one third of cars, with Japan in third place and rising fast. Cement making was also a big fuel consumer during the rich countries' first stages of industrial growth, to build roads, bridges and buildings.[17]

Industrial agriculture, which originated in the USA and spread across temperate zones during the boom, became another significant consumer of fossil fuels. Systems were put in place that used oil and gas as raw material for fertiliser, and to power farm machinery, vehicles to transport food, and refrigerators. In the 1949/50 season, world consumption of fertilisers was estimated at 12.46 million tonnes (mt); after 17 boom years, in the 1966/67 season, it had risen to 48.3 mt.

The concentration of industrial agriculture was reflected in the tractor fleet. In 1950, the world's farms had an estimated 6.16 million tractors, of which 5.21 million were in North America and Europe. In 1966 there were 13.81 million tractors, of which 9.96 million were in North America and Europe.[18] By contrast, in 1951 India had 8,500 tractors and nearly 80 million draught animals, and things had hardly changed by the late 1960s. In China, the capacity of motors in agriculture (including tractors, pumps and other equipment) rose fivefold between 1965 and 1974, but was still then estimated to be below that provided by animal and human muscles. The export of fossil-fuel-intensive agricultural products (wheat and other cereals in particular), from the USA and other rich countries to developing nations, was an important characteristic of economic domination.[19]

Industrial agriculture was credited with huge productivity gains, and these helped to ensure sufficient food was available for a rapidly growing population. The rich world began shifting to meat-based diets, and this improved nutrition – although the extent to which it did so, and the accompanying disadvantages, are subjects of contention. The rate at which fossil fuels went into agriculture rose even faster than output. Between 1945 and 1970, crop yields rose by 138 per cent, while energy inputs rose by 213 per cent, due to mechanisation, the ramp-up in fertiliser use, and the drying and transport of grain.[20] In 1980, 30 times more fertiliser was being applied per hectare of US cornfield than in 1945, while corn yields per hectare had risen only four times. Such fossil-fuel-driven productivity improvements essentially reached their limits in the 1980s, after which agricultural yields rose less rapidly.[21]

Roads and cars

A striking difference between the USA and other rich countries was the level of car ownership, and the way that the car had become inscribed in the USA's economy, society and culture. In 1950, US citizens drove 40 million cars, about 75 per cent of the world total. In Europe, only the UK and France came anywhere near US levels of car ownership, with 2.25 million and 2 million respectively, but large swathes of both were still reliant on horses and carts. Spain had just 89,000 cars, one for every 314 people. The situation outside the rich world was exemplified by India, which in 1960 had 263,000 cars (one for every 1650 people), compared to the USA's 61.7 million cars (one for every 2.9 people).[22]

The move to mass car ownership had begun in the USA before the Second World War, together with suburbanisation, the decline of public transport and government support for road building. (See pp. 21–3.) All these trends accelerated during the post-war boom. This was not spontaneous: an alliance of government and the big three car manufacturers (General Motors, Ford and Chrysler), just as significant as the alliance of government and multinationals in oil production, helped to make it happen. The 1956 Federal Aid Highway Act – adopted after vigorous campaigning by the National Highway Users Conference, a lobbying group supported by GM – provided for the 66,000 km Interstate Highway System, the largest single construction project in history. Huge state investment in federal highways – $70 billion in the 15 years 1956–1970 – was justified, in part, with reference to national security. This sum dwarfed other infrastructure budgets: investment in rail transit in the same period was less than 1.5 per cent of that sum, $795 million; even the entire Marshall Plan (US aid for post-war reconstruction in Europe) cost only $17 billion. A corollary of highways expansion was a continued onslaught on railways and public transit. In 1950, there were 17.2 billion journeys on buses and trolleys; by the early 1970s, fewer than 7 billion per year. In the decade from 1954, 200 transit companies went out of business. In Europe, by contrast, electric streetcars continued to be run as a public service, often at a loss.[23]

The waste inherent in the manufacture and use of cars was well understood, long before the 1973 oil shock forced it to the centre of public attention. In 1962, the cost of annual model changes was estimated at more than a quarter of the purchase price of each new car, and about one-fifth of the petrol costs its owner would pay. While technology allowed for greater fuel efficiency, the actual fuel efficiency of new cars fell as they got heavier, from 16.4 miles per gallon (mpg) in 1949 to 15.2 mpg in 1961. And the cost to motorists of model changes was at least $5 billion/year in 1956–60. Building on this research, the Marxist economists Paul Baran and Paul Sweezy estimated that the real cost of production and distribution of a functional 1949 model car in the early 1960s would have been about a third of the current sales prices, and that the extra paid by millions of drivers was probably equal to 2.5 per cent of the Gross National Product. As soon as the 1973 oil shock hit, the US state of Illinois commissioned researchers to define the technical potential for reducing cars' energy costs: the answer was that one-third of the energy used to make cars could be saved, and that if cars' lifetimes were extended from 10 to 30 years, the saving would be 96 per cent.[24]

The highway-based transport system – in the USA, Canada and Australia, and then elsewhere – went with the 'automobile city', typically spread out with a 50 km radius and low population density of 1-2000 people per square kilometre. Urban planners set out alternatives. 'Expressways eviscerate cities', wrote Jane Jacobs; 'the more space that is provided [for] cars in cities, the greater becomes the need for use of cars, and hence for still more space for them.' The urban activism she helped inspire constrained highway construction in the 1970s and 1980s.[25]

Suburban housing – single-family homes with their own garden spaces – was a key feature of car-based cities. In the USA in the early 1950s, more than 1 million houses were built per year. Consumer credit was part of the deal: the Federal Housing Administration offered mortgages with 25- or 30-year terms. Homeowners, as a proportion of the population, rose from 44 per cent in 1940 to 62 per cent in 1960. For the first time ever, large numbers of people lived lives that were almost impossible without a car. By 1960, nearly two-thirds of the working population drove to work, and in 1969 their average journey was estimated at 16 km.[26]

By 1970 some European countries had made the transition to mass car ownership – the UK then had 11.5 million cars, France 12 million and Italy 10 million – but the USA had meanwhile risen to another level. It had 90 million cars, enough to carry its entire population, sitting only in the front seats.[27]

Rich world household consumption

Households, which accounted for around one fifth of rich-world fossil fuel consumption during the boom, used energy for four main purposes: to heat homes; to heat water e.g. for washing and bathing; to cook; and for electrical appliances. Heating took the lion's share: 80 per cent in western European countries such as France, west Germany and Italy; less than 70 per cent in the USA and the UK; and 40 per cent in Japan. The energy researchers Bertrand Chateau and Bruno Lapillonne found that consumption growth accelerated in the 1960s, but patterns were 'highly differentiated' between countries.[28] The share used by electrical appliances would grow over time.

In Europe, there was large-scale migration from the countryside to the cities. Between 1951 and 1971, the population of cities such as Athens and Milan grew by nearly half; Belgrade's, Sofia's and Bucharest's doubled. By and large, the countryside stuck with coal – even by 1975, only a quarter of French farmhouses had central heating – but in cities, central heating became the norm. It was retrofitted to old houses and often came as standard in new ones, whatever the climatic conditions. Rising incomes, home ownership and low-interest home improvement loans, hastened the changes. In France, 90 per cent of homes built after 1962 were fitted with central heating; in the Netherlands, it went into all homes built after 1965.[29] Fuel consumption levels depended largely on how energy-efficient buildings were, and what demands regulators placed on construction firms. There was little or no insulation. In the 1980s researchers found that for every 1 unit of heat used, per square metre of floor area, in a new US home, an old US home used 1.6 units, a new home in Sweden 0.65 units, and energy-efficient demonstration homes 0.15–0.83 units.[30] With central heating came the widespread heating of unused rooms, a persistent cause of fuel waste to this day.

Ever-higher inside temperatures in winter became 'considered by many as a condition for comfort and well-being', in Chateau and Lapillonne's judgment. So, too, did hot water – in the 1950s 'a symbol of comfort and luxury which concerned [only] the well-off', but by the 1960s and 1970s becoming 'rapidly widespread'. In the USA and the UK, most homes had hot water by the 1950s; in continental Europe, the proportion of homes with hot water rose from 30–40 per cent in 1960 to 75–95 per cent by 1975.[31]

In the USA, air conditioning became widespread. The doubts of homeowners and office managers – many of whom regarded it as a luxury – were overridden by the construction industry, who were convinced by air conditioning manufacturers to install it as standard. Air-conditioning systems made possible urban growth and industrial production, along typical rich-country lines, in the tropical climate of the southern states. In the 1960s, air conditioners were a major cause of the reversal of the traditional trend of out-migration from the south to other parts of the USA.[32]

During the boom, almost every US household acquired the first generation of electrical appliances – radios, refrigerators, washing machines and vacuum cleaners – and Europe, Japan and Australia began to follow. In the USA in 1950, 80 per cent of households had a refrigerator and 54 per cent a vacuum cleaner; by 1970, those levels were 99 per cent and 92 per cent. In the UK, a household survey in the late 1950s found that the first domestic appliance acquired was usually a cooker; the second a vacuum cleaner; the third a washing machine or refrigerator. Continental western European households were relatively slow to follow. In 1957, only a small minority of them had a refrigerator – 2 per cent in Italy, 12 per cent in West Germany – mainly because incomes did not allow people to stock up on food sufficiently to make buying a refrigerator worthwhile. But by 1974 the transition was complete: 94 per cent of Italian homes and 93 per cent in West Germany had a refrigerator. Washing machines took longer to catch on, partly because many European households were still without running water in the mid-1950s.[33]

Sociologists' research on the order in which households acquired appliances showed that everywhere they started with those that eased the most physically demanding tasks. In the Midwestern USA in the 1960s, they first purchased an electric iron, and then a car. Refrigerators – that potentially save time spent storing, collecting or purchasing food – were a greater priority for rural households than for urban ones, and more important than washing machines in town and country alike. In Israel in 1965, households were likely to purchase a radio first, and then a cooker, refrigerator and washing machine; a survey of surveys found that that was a rough guide to behaviour internationally. By the time appliances became widespread in India and China in the 2000s, the order had shifted. Households often bought sewing machines – probably reflecting a high level of paid domestic labour – followed by watches, bicycles and radios. Only after these did they buy TVs, refrigerators, electric fans and washing machines.[34]

The refrigerator's history provides insights into how the evolution of appliances affects fuel consumption levels. Manufacturers persistently increased refrigerator sizes, and convinced households of the need for larger ones. In the USA, average sizes rose from 0.1–0.2 cubic metres in the 1930s to 0.34–0.4 cubic metres in the 1960s, and 0.5–0.74 cubic metres in the 2000s. Refrigerators were about half that size elsewhere – in 1998, 0.23 cubic metres in Russia and 0.27 cubic metres in Europe, on average, compared to 0.5 cubic metres in the USA. The energy efficiency of new refrigerators fell steadily until the 1970s, but then (fortunately, and unlike cars) improved dramatically, more than tripling between 1972 and 1996 – although energy savings from efficiency were partly cancelled out, to the extent that refrigerator sizes increased. In the USA, freezers – either free-standing or as part of refrigerators – took off in the 1960s; their success required not only cultural acceptance of frozen food, but also a cold chain including warehouses, railcars and trucks.[35]

Outside the rich world: the Soviet Union

The Soviet Union, which emerged from the Second World War as the second superpower after the USA, in the 1950s and 1960s made the transition from a mostly rural to a mostly urban society, and from an energy system based on coal and biomass to one heavily reliant on oil and gas. Its energy system, and especially its huge electrical network, was seen as an alternative, non-capitalist development model by national elites across the developing world. Soviet energy consumption rose far faster than in the rich countries – 10.5 per cent/year in 1925–50 and 7.5 per cent/year in the post-war years – to support the expansion of heavy industry.[36] In the post-war period sections of the urban population for the first time acquired central heating and running water, but there were few electrical appliances, and even fewer cars, before the 1970s.

Immediately after the war, the USSR relied heavily on its already-developed coal, peat and wood industries, and planners were reluctant to risk resources on potentially high cost oil and gas exploration. But when the decision was taken in the mid-1960s to develop Siberia's oil and gas reserves, things changed quickly. In 1959, solid fuels (mostly coal, peat and firewood) accounted for 65 per cent of primary fuel consumption; by 1975 their share was 32 per cent and oil and gas's share was 64 per cent. The replacement of coal by gas in the electricity and heat sector was especially rapid.[37]

The electricity system was essentially two-tier. Centralised grids came together in European Russia in the late 1950s and central Siberia in the early

1960s, but only supplied cities and industrial centres. Other consumers, and in particular the state-owned collective farms on which most of the rural population lived, had to fend for themselves – mostly with small, unconnected local stations. (See p. 112.) The primary consumer of electricity, and of all fossil fuels, was the prodigious Soviet industrial sector, which in the mid-1970s used more than two-thirds of all electricity, compared to 40 per cent in the USA. Household consumers' share was 14 per cent, compared to 58 per cent in the USA. Metallurgy, chemicals, machine building and the fossil fuel industries themselves were the largest users. Rural households' meagre electricity consumption was supplemented by biomass. Even in the late 1970s, more than a quarter of the fuel used by households was firewood and peat, much of it gathered by consumers themselves in the countryside.[38]

The Soviet Union urbanised rapidly. Between 1959 and 1968 it provided new housing to 87 million people. Space was painfully short, and remained so until the 1990s: the average space per person was 8.1 square metres in 1958, rising to 11.2 square metres in 1971, compared to 20.8 square metres in East Germany. The ubiquitous urban apartment blocks were supplied by technologically cutting-edge district heating systems, which used surplus heat from electricity generation, or from industry via boilers, or from autonomous coal- or gas-fired boilers. (See p. 30.)[39] Urban transport was public, and cars a privilege of the elite, until the end of the Soviet period: in 1970 car densities in the Soviet bloc ranged from one per 150 citizens in the USSR to one per 15 citizens in East Germany.[40]

Outside the rich world: Asia, Africa and Latin America

In the 1950s–1960s, most energy systems outside the rich world – on which more than half of the world's population relied – were non-commercial systems dominated by biomass. (See pp. 40–1.) Fossil-fuel-based commercial systems became significant in a minority of developing countries. There was a sharp differentiation between developing countries. Research on 'the non-communist developing world' (excluding China and Vietnam), showed that between 1960 and 1972, commercial energy consumption more than doubled, but three-quarters of it was concentrated in just 16 countries: Indonesia, Iran, Mexico, Venezuela, Colombia and Argentina (all significant oil producers); Brazil, Egypt, India, Pakistan and Turkey (all with large populations and economies), and Chile, the Philippines, Taiwan, Thailand and South Korea. Consumption levels in these countries were far above those in most of Africa and much of Asia, but far below those in the rich world. In 1970, non-commercial fuels still made up 90 per cent of India's energy balance and 75 per cent of Indonesia's.[41]

The non-commercial energy system in rural areas outside the rich world used few, or no, fossil fuels, and usually went with subsistence agriculture characterised by extremely low efficiency, reliance on 'distressingly hard and unproductive labour by human beings and their animals', and a heavy burden on wood resources, the development researcher Arjun Makhijani wrote. Wood fuel, dung and crop residues were used for cooking, heating, lighting, water heating and ironing clothes. During the 1960s and 1970s, biomass was supplemented by some coal, oil and electricity – but these were used almost exclusively by the richest one-fifth of the rural population. In 1971 the World Bank estimated that 4 per cent of the rural population in sub-Saharan Africa, 15 per cent in Asia and 23 per cent in Latin America were served by electricity. Those figures referred to villages, though, and only a small proportion of the populations of electrified villages actually had hook-ups. (See pp. 109–10.)[42]

The urban poor in developing countries had some access to commercial fossil fuel energy but often no means of paying for it. In the countryside, wood might be substituted by other biomass when it was not available, and cash crops could be traded for charcoal, paraffin or LPG for cooking. But in towns, there were few opportunities for collecting free fuel e.g. from forests, and alternatives often cost money. Typically, charcoal was the main fuel for the urban poor, but if wood was available in the form of sawmill waste, discarded packing cases, splits, or offcuts, people would collect or purchase it. Only middle- and high-income families started to use paraffin and bottled gas. In Bangalore, India, researchers found in the 1970s that wood was the main fuel in just under half of households; some could access coke and electricity. In Mexico City, kerosene and electricity dominated, except for the lowest income group, which often fell back on wood and coke.[43]

In the post-war period, developing country governments often prioritised industrialisation and urban electrification. In India, after independence from the British Empire (1947), the government invested heavily in infrastructure and capital goods industries. It sought self-sufficiency in steel, cement, fertilisers, metals and machinery. In 1955, the lion's share of commercial energy (which was mostly coal, coke and oil products) went to industry (46 per cent) and transport (31 per cent), while households relied almost entirely on biomass. Industry also consumed most (72 per cent) of electricity.[44]

The provision of energy resources and fertilisers to expand agricultural output was also seen as crucial by governments and development specialists. In India, the Philippines and Afghanistan especially, US aid programmes encouraged fossil-fuel-intensive farming methods, which went together with the high-yield seeds and associated technological know-how that US politicians and academics saw as weapons of the Cold War in Asia. Between 1950 and 1970, India's food production doubled and its population rose by 50 per cent. In spite of terrible droughts in the mid-1960s, per capita food output grew, but in an unequal way, Makhijani wrote. Punjab and other richer states experienced 'spectacular' agricultural production growth, thanks to high-yielding varieties of wheat and greater shares of fertilisers and pesticides, while other states were constrained by lack of irrigation. Rural electrification was also unequal: in 1974, in Tamil Nadu 97 per cent of villages had access, and in Punjab 55 per cent, while only 15 per cent of villages in Bihar, and 3 per cent in Assam, did.[45]

In China, the Communist Party's accession to power in 1949 was followed by a period of post-conflict reconstruction, and then by a series of economic plans in which rural electrification and agricultural output growth figured consistently. In the late 1950s the disastrous 'great leap forward' attempted to force forward industrialisation; economic slump and famine followed. In the 1960s, rural development policy paid more attention to agriculture. Huge numbers of small coal mines (110,000 at one count in the 1960s), and small hydro stations, were commissioned. Although urban areas and industry were usually first in line for grid connections, some fossil fuels and electricity began to be used on farms.[46]

Domestic labour and household consumption: interpretive issues

The post-war surge in rich-world household energy consumption transformed domestic labour, but not straightforwardly. Homes were electrified and supplied with gas. Appliances, in the first place gas cookers, refrigerators, washing machines and vacuum cleaners, were acquired by hundreds of millions of people. Historians collectively have concluded that (i) the character of domestic labour was overturned by these appliances, (ii) they did less than might have been expected to lift the burden of that labour from housewives, and (iii) specifically, the hours women spend doing housework have not fallen very far as a result. Technology changed people's lives, but only within constraints shaped by society, economics and culture.

Ruth Schwartz Cowan, the historian of technology, coined the term 'industrialisation of the home', to emphasise that modern housework depends on non-human energy sources just as industrial work does, and that it is integrated into wider economic and social networks, as industrial work is. She showed how, in the USA in the early twentieth century, tasks such as food preparation moved out of the home, thanks to the food processing industry and refrigeration. But that time previously spent by housewives on food preparation was spent, instead, on motorised shopping trips. Appliances such as washing machines and vacuum cleaners were huge labour savers – but this labour was more likely to be that of domestic servants (whose numbers fell steeply), men, or children, than housewives'. By the post-war years, when such technologies were ubiquitous, 'there is more work for a mother to do in a modern home because there is no-one left to help her with it'. Moreover, household technological systems 'were built on the assumption that a full-time housewife would be operating them' – and could not be changed rapidly when full-time housewives began to disappear, usually by joining the labour force.[47]

The other great change that culminated in the USA during the post-war boom, and elsewhere subsequently, was in what household work produced. Standards of sanitation and cleanliness rose; assumptions about childcare changed; and so did ideas about nutrition (not always for the better). Changes in housework were analogous to changes in industry. There, too, new types of energy consumption and new technology increased the productivity of labour, but – due to the social and economic system under which work was being done – never realised their potential in terms of reducing working hours.

Because housework is unpaid, and often regarded (and not only by economists) as somehow inferior, it is more difficult to measure. The historian Joann Vanek compared housework in the USA in the 1920s and the 1970s and found that, on average, for women who did not have paid jobs outside the home, the hours worked were the same. But the nature of the work had changed: less time was spent on producing food and making clothes, more on shopping and childcare. Valerie Ramey concluded from a study of the USA between 1900 and 2005 that average hours of housework by women, including those who worked outside the home, fell by a little more than a third over the twentieth century (from 46.8 to 29.3), with most of the reduction coming after 1965; that this was not matched by an increase of housework by men; and that for many women drudgery was reduced, but not hours.[48]

An important example from outside the rich world is that of the USSR, where most urban housewives had no washing machines, vacuum cleaners or even refrigerators until the 1970s – and, unlike urban women in capitalist countries, had to queue, sometimes for hours, to buy food. Moreover, a much greater proportion of them worked outside the home. The result, combined with the blunt refusal by many men to share housework equally, was the notorious 'double burden' of work and housework. In the 1960s, 'anyone who discussed the problem saw the solution in the provision of public services and labour-saving appliances [which, however, were not yet forthcoming], rather than in the organisation of society and the attitudes of men that it engendered', the labour historian Donald Filtzer lamented.[49]

A research project in Kerala, India, in the 1990s revealed similar social and economic contexts, but – given easier access to appliances – different outcomes. Increased educational opportunities had led to more women working outside the home; this produced 'tremendous time pressure on wives'; this in turn stimulated purchases of washing machines and refrigerators. Families in which women did not work were far less likely to have a washing machine than those where they did.[50]

Washing machines, refrigerators and other labour-saving appliances were not, then, the 'engines of liberation' some economists made them out to be.[51]

Neither, though, were they dispensable luxuries. 'Those social critics who disparaged [...] expenditures [on electricity, running water, bathrooms, gas ranges, telephones and cars] were uniformly members of the more comfortable classes', Cowan wrote. Such moralisers were more likely to be American Studies specialists than housewives, Stanley Lebergott, the historian of consumption, pointed out, pouring scorn on those who wrote critically about women who bought refrigerators and canned foods.[52]

6
The 1970s: crises and oil price shocks

In the early 1970s, the post-war boom gave way to an era of economic and political instability. The rich countries experienced recessions in 1974-5 and again in 1980-82. The USA, which had emerged from the Second World War so completely dominant, went into relative decline. Its grip on world financial architecture was loosened in 1971, when the dollar was delinked from gold, and it lost position militarily, with its humiliating withdrawal from Vietnam in 1975. The early 1970s were also a turning point for fossil fuel markets. Rapid oil and gas consumption growth, characteristic of the boom, accelerated towards its end: OECD consumption of oil rose by 29 per cent in the three years to 1970, and a further 20 per cent in the three years to 1973; of gas, by 14 per cent and 30 per cent, although their coal consumption fell slightly.[1] In 1970-73, as oil became a sellers' market, the rich countries became vulnerable to a drastic rebalancing of it in producing nations' favour.

From 1970, members of the Organisation of the Petroleum Exporting Countries (OPEC) took control of production away from the mainly US-based international oil companies (IOCs), and raised prices. Attempts to manage the rebalancing collapsed in October 1973. At the same time, war broke out between Israel and a coalition of Arab states led by Egypt and Syria; in response to western support for Israel, Middle Eastern producing nations placed an embargo on oil sales to the USA. The first oil price shock followed: a barrel of crude, which had cost around $1.80 throughout the 1960s, and $3.29 in 1973, rose to $11.58 in 1974. It rose gently until 1979, when the Iranian revolution, and the transition from contract pricing to spot market pricing, produced a second price shock: the average price in 1980 was $36.83.

It is often said that 1973 brought an 'energy crisis'. This mixes up at least three distinct phenomena. First, there was a real crisis in developing countries that were dependent on oil imports. After the price increase, Brazil, Argentina, Chile, India, Turkey and a swathe of African countries ran up tens of billions of dollars worth of debts, to pay for oil supplies without which they risked economic and social collapse. Second, there was a temporary and partly illusory 'crisis' in rich countries, involving sharp price increases and, briefly, shortages of petrol (such as those experienced continually outside the rich world). Third, there was a crisis of perception and policy. Among rich-world political elites and populations alike, the boom-time abundance of fossil fuels had given rise to a sense that cheap energy sources would always be available, and always under those elites' control. The OPEC price increases destroyed such assumptions. Reactions ranged from nationalistic calls for 'energy independence' on one hand, to consideration of environmentalist and energy conservation policies on the other.

The oil price increases led in the rich countries to energy efficiency gains, particularly in industry, and to some energy conservation. Some mostly short-lived efforts were made to develop renewable energy sources. OECD countries' oil use fell; coal, gas and nuclear power gained ground. Only when the economy went into recession did overall OECD commercial energy consumption fall. Many opportunities for deep-seated change in consumption practices were missed. The USA, in particular, sought to insulate energy-intensive systems and practices from change, e.g. by subsidising petrol for car drivers.

The 1973 oil price shock

The causes of the 1973 oil price shock lay above all in the one-sided character of the international oil supply and trading system. In the 1940s–1950s, the 'seven sisters' dominated oil production internationally, controlled oil transport and trading, and with US government support, set prices at a level that ensured good margins. The USA covered almost all its oil needs from domestic output, and had sufficient spare capacity to help out allies in the event of supply disruptions – as it did during the 1956 Suez crisis, and during the 1967 Six Day War between Israel and Arab nations, when OPEC countries attempted an embargo. In the 1960s, this system began to change. First, oil production shifted geographically towards the Middle East. (See p. 80.) Second, competition opened up, as supplies not controlled by the 'seven sisters' became available; prices fell in response.[2] Third, by the early 1970s, the USA was becoming much more import dependent, while its own oil output, from relatively costly fields, had peaked. Oil imports to the USA more than doubled between 1970 and 1973,[3] and while it was still less dependent on imports than western Europe and Japan, it no longer had spare oil to export to those countries if supply was disrupted.[4]

As oil demand soared, OPEC countries sought to take control of the oilfields on their territory away from the IOCs, and to increase prices in a co-ordinated way. The rise of anti-imperialist movements in the Middle East hastened the process. In 1970, Muammar Qaddafi took power in Libya; his government, and Venezuela's, initiated demands that the IOCs renegotiate prices. Then Algeria (1971), Iraq (1972) and Libya (1973) nationalised oil assets, and Saudi Arabia (1972), Abu Dhabi (1972), Kuwait (1973) and Qatar (1973) struck agreements under which IOCs ceded control of assets to national oil companies. All told, between 1970 and 1979, the share of producing nations' assets (excluding North America and the Soviet bloc) owned by national oil companies rose from 10 per cent to 79 per cent, and the IOCs' shares fell by roughly the same extent. In 1971, at meetings in Tehran and Tripoli between OPEC producers and IOCs, a series of staged price rises were agreed – but OPEC sensed it could go further. In 1973, talks reopened, but collapsed amid political tension generated by the Arab-Israeli war. In protest at western support for Israel, OPEC cut oil exports by 10 per cent, and later briefly by 25 per cent, and placed a selective embargo on the USA and some other importers. The embargo had little effect, but by January 1974 OPEC had been able unilaterally to raise the sales price of crude oil nearly fourfold, to $11.65/barrel.[5]

The oil price increase exacerbated an economic crisis that had been mounting since the late 1960s. In the rich countries, the long boom had produced over accumulation of capital relative to labour, and, as demand for labour grew, there was a wave of strikes across western Europe in 1968–9. Profit rates fell; inflation soared; commodity prices rose particularly rapidly; in 1971 the dollar was devalued. In 1974, economies crashed. Rich countries' trade deficits deteriorated, due largely to greater oil import bills. Investment collapsed. Unemployment in the OECD rose from 11 million in 1973 to more than 18 million in 1976. Rising oil prices aggravated consumer price inflation, which reached 30–35 per cent in Japan, 24 per cent in the UK, and double figures more widely. Money drained to the oil producing countries: OPEC's exports, which in 1970 had been valued at $15 billion, were in 1978 valued at $136 billion. The flow of funds out of the rich countries exacerbated both industrial stagnation and inflation – a combination named 'stagflation'.[6]

Politically, the rich countries were divided on how to respond to OPEC's actions. The USA had its own oil, while others did not; the western powers had different strategic interests in the Middle East. At the USA's urging, the IEA was set up in 1974 to coordinate responses to OPEC, but for that purpose it remained largely ineffective.

Between 1973 and 1975, consumption of oil, gas and coal by OECD countries fell for the first time since the boom began. Consumption growth resumed in the OECD in 1975, albeit at a slower rate than in the Global South and in the Soviet bloc. Overall, oil demand rose strongly enough to lift prices, from $11.53/barrel in 1975 to $14.02/barrel in 1978. In 1979, the revolution in Iran and a national oil workers' strike halted production for more than two months and signalled a new political crisis in the Middle East. It also marked a new transition in oil trading: greater volumes were sold on the spot market and OPEC lost much of its price-setting power. Markets are vulnerable to panic, and that played a part in driving oil prices to very high levels – $31.61/barrel in 1979, and a peak of $36.83/barrel in 1980 – before they settled at $27–35, before falling to $14.43 in 1986.[7]

OECD oil consumption fell after 1973, recovered to a new peak in 1979, and then fell again until the mid-1980s. Through the two oil price shocks, OECD coal consumption rose substantially, as the high price of oil imports encouraged use of coal as a substitute. Outside the OECD, oil consumption rose consistently through the whole period. This was due, firstly, to growing oil use in the Soviet bloc. Natural gas consumption also continued to rise there. The Soviet bloc countries were making a shift away from a predominantly coal-based economy that rich western countries had largely made during the post-war boom. In the Global South, consumption of all fossil fuels rose through both the oil price shocks. Many of them took drastic measures to reduce dependence on imported oil, but the net effect of this on consumption levels was outweighed by the impact of industrialisation, electrification and urbanisation. The Global South's consumption per head moved from an extremely low level to a very low level.[8] (See Table 8.)

Oil consumption patterns changed as a result of the two oil shocks. (See Table 9, p. 98.) The comparison between the USA and Japan, two of the richest countries, is revealing. The USA had become partly dependent on imports, but still produced most of the oil it consumed; thanks to a two-tier pricing system (discussed below), its industrial consumers could buy oil well below world market prices (about 40 per cent below in early 1979, for example). Japan, by contrast, was almost entirely dependent on oil imports and had to borrow heavily to pay for them. It increased use of coal and nuclear energy substantially; oil, as a proportion of Japan's commercial energy supply, fell from more than three-quarters in 1973 to just over half in 1985. Oil's share of the USA's total commercial energy supply fell much less dramatically, from 46 per cent to 40 per cent – and the total commercial energy supply rose substantially. Argentina and Brazil are examples of developing countries that went into the 1970s with big oil import bills: between 1973 and 1985, oil's share of commercial energy supply fell almost as sharply as in Japan – from 73 per cent to 56 per cent in Argentina, and from 71 per cent to 50 per cent in Brazil.[9]

The next three sections cover in more detail, respectively, changes in OECD fuel consumption patterns through the two oil shocks; the changes in perceptions and policies that the oil shocks caused; and the shocks' effect outside the rich world.

Impact on the rich countries

The oil shock pushed up prices of all fossil fuels in OECD countries. Between 1960 and 1973, the price of oil had fallen in real terms, and so 'producers of other forms of energy were forced to keep prices, and consequently production costs, low' – but after 1973, the prices of other fuels were 'rapidly adjusted upwards', especially for industrial consumers, a survey by the IEA found. The price of

Table 8 Commercial energy consumption through two oil shocks

Million tonnes of oil equivalent		1972	1973	1974	1975	1976	1977	1978	1979	1980	1981	1982	1983	1984	1985
OECD countries	Total	3725	3932	3872	3790	3995	4084	4147	4256	4147	4050	3940	3937	4123	4191
	Oil	1908	2049	1964	1905	2032	2079	2099	2110	1966	1848	1749	1718	1765	1745
	Gas	748	768	772	739	767	770	788	827	821	806	771	749	804	803
	Coal	825	861	852	841	884	903	894	945	974	983	976	992	1038	1081
	Nuclear	32	42	54	74	87	107	125	127	140	164	178	198	236	283
	Hydro & other rnwbls	212	212	231	232	226	225	240	248	247	249	266	279	279	279
Non-OECD countries	Total	1693	1791	1875	1984	2095	2223	2355	2463	2487	2536	2621	2712	2847	2981
	Oil	663	723	768	796	843	892	953	992	1016	1023	1030	1038	1052	1071
	Gas	258	284	305	332	367	397	424	462	475	502	541	579	635	686
	Coal	689	696	703	749	769	806	838	856	830	835	869	902	943	989
	Nuclear	3	4	6	8	11	14	15	18	21	25	30	35	45	53
	Hydro & other rnwbls	81	85	94	98	105	115	125	135	145	150	151	158	172	181

Note: figures have been rounded to nearest mtoe. Source: *BP Statistical Review of World Energy*

Table 9 Oil consumption in selected countries, 1965–85

In millions of tonnes, and as proportion of total commercial primary energy consumption (tpec)

		1965	1970	First oil shock			Second oil shock			
				1973	1974	1975	1980	1981	1982	1985
US	mt	551.3	705.7	833.1	799.7	783.5	788.1	734.3	694.1	709.9
	% of tpec	42.9%	43.4%	46.0%	45.3%	45.6%	43.5%	41.7%	41.2%	40.4%
Germany	mt	86.3	138.7	162.2	147.0	142.6	147.3	133.4	125.9	126.3
	% of tpec	34.0%	45.2%	47.9%	44.2%	44.7%	41.1%	38.4%	37.6%	35.0%
Japan	mt	87.9	199.2	269.2	259.0	244.1	237.8	223.8	209.5	207.6
	% of tpec	57.1%	71.1%	77.3%	74.5%	73.7%	66.8%	63.9%	62.0%	56.2%
Argentina	mt	22.0	22.1	23.7	23.9	22.4	24.3	22.4	21.5	18.7
	% of tpec	81.9%	76.6%	72.7%	69.9%	67.4%	61.4%	58.9%	55.0%	47.1%
Brazil	mt	14.8	25.0	39.0	42.0	43.3	54.5	51.8	53.1	54.3
	% of tpec	67.3%	68.3%	70.9%	69.9%	68.3%	60.0%	58.6%	57.2%	50.1%
India	mt	12.6	19.5	23.3	22.8	23.3	31.6	34.0	35.4	43.3
	% of tpec	24.0%	30.1%	32.4%	29.9%	28.5%	30.9%	30.0%	31.4%	32.4%
Price of oil, $/barrel		1.80	1.80	$3.29	$11.58	$11.53	$36.83	$35.93	$32.97	$27.56

Source: BP Statistical Review of World Energy

natural gas, which often competed with oil products such as fuel oil and heating oil, 'increased very steeply [...] though from an initially lower level'. The US and other governments encouraged the use of coal as an alternative to oil, helping to produce a new generation of coal-fired power stations. Coal mining companies, whose profits had been depressed for more than a decade before 1973, were suddenly able to make money again. Electricity prices rose less rapidly, partly because the cost of nuclear and hydro-produced electricity did not change.[10]

The prices of oil products – petrol for cars, heating oil and fuel oil, and other refinery outputs from asphalt and tar at the bottom end to kerosene and aviation fuel at the top – were (and are) made up mainly of the crude oil price, plus refinery and transportation margins, plus taxation. In most OECD countries, between 1973 and 1980, 'motorists appear to have been "insulated" to a large extent from oil prices rises, compared to other classes of consumers', the IEA found. Between 1973 and 1979, posted prices of crude rose by 680 per cent, while retail petrol prices rose by 115 per cent in the USA, 32 per cent in West Germany, 138 per cent in France and 226 per cent in the UK. The USA, champion of car culture, used low tax rates to keep petrol prices far lower than in other rich countries. Between 1973 and 1980 the level of petrol taxation, as a proportion of the ex-tax price, fell from 38 per cent to 13 per cent in the USA, compared to a drop from 239 per cent to 111 per cent in West Germany and 160 per cent to 46 per cent in the UK. (For comparative petrol and electricity prices in rich countries in 1980, see Table 14, p. 210.)[11]

OECD governments' concessions to motorists were reflected in petrol consumption, which continued to grow after 1973, albeit at a slower rate. Energy consumption in the transport sector (mostly petrol for cars) grew on average by 5 per cent per year in 1960–73, and a lower but still healthy 2.1 per cent per year in 1974–80. Average petrol consumption per car fell on average by 3 per cent per year in 1974–80, but that was mainly thanks to drivers in, for example, Japan and Italy, using smaller, more efficient cars; the reduction in the USA and Canada was modest by comparison. In the USA, the distance driven by the average car fell during the 1970s, and its fuel efficiency rose by 13 per cent – but this was outweighed by a big increase in the number of cars. Between 1969 and 1983 the number of kilometres driven by the average household rose by 29 per cent; there were 39 per cent more shopping trips; and the distance travelled on these trips rose by 20 per cent on average. Over the long term, the oil shocks hardly dented the remorseless rise of fuel consumption by cars. In the quarter century 1973–98, growth in passenger transport (mostly private cars) was the biggest contributor to rising OECD oil demand: its share, in the 11 largest OECD economies, rose from 38 per cent to 53 per cent. Car ownership rates doubled, and fuel intensity fell, but the effect was cancelled out by the trends to longer journeys and heavier, more gas-guzzling cars.[12]

In industry, the oil price shocks had two effects. First, firms made energy efficiency improvements, in order to cut fuel costs. Second, there were structural changes – the start of a trend that would continue into the twenty-first century – whereby companies exported energy-intensive processes to the Global South where labour was cheaper, and in the rich countries focused on more profitable fabrication and finishing processes. In the seven largest OECD economies, the energy intensity of industry fell after 1973, most sharply in the UK, USA and Japan. Changes in the industrial structure, more than efficiency improvements in particular factories, were probably 'the principal reason' for the trend, the IEA concluded. Between 1972 and 1979, in the US chemical industry, the amount of value added rose by 37 per cent, while the amounts of feedstock and fuel used rose, respectively, by 26 per cent and just 4 per cent. One group of researchers cautioned that, in US industry, most energy conservation was simply economically rational cost cutting by saving, or switching, fuels.[13]

Research of 32 rich countries between 1970 and 1987 showed that western European countries such as France, West Germany and the UK expanded their economies more rapidly than they expanded energy-intensive production of steel and cement. But the opposite was true in southern Europe (e.g. Greece and Portugal) and Eastern Europe (e.g. Bulgaria and the USSR). The contrast was striking in Germany: in the west, the service sector contributed to economic growth at 'relatively little cost in terms of energy and materials'; in the east, the 'materials economy', including polluting heavy industry, was the main source of economic growth.[14]

Overall, research of the oil shocks' effect on rich countries' fuel consumption points to four conclusions. First, the shocks gave a new lease of life to coal, which was substituted for oil products where that could cut costs. Second, the shocks triggered a big shift in rich-country industry from energy-intensive basic materials processing towards fabrication and finishing. Third, the conclusion drawn by the IEA from its survey of the eleven largest OECD economies, which showed an overall fall in industry's energy intensity of 4 per cent per year in 1973–82: 'Before 1973, energy prices were generally low, so when the price hikes kicked in after 1973 there was ample room for improving energy efficiency as a response. As prices fell after 1985, the incentive for maintaining energy savings rates weakened.' Fourth, the energy efficiency improvements in industry, stimulated by the need to cut costs, were not replicated in the transport sector, mainly because governments protected motorists from the impact of the oil shock, by reducing taxes on petrol.[15]

Crises of perception and policy

The oil price shocks, and perceptions of an 'energy crisis', for the first time prompted serious consideration of energy conservation in rich-country political elites. Usually, though, these discussions took second place behind those about building stockpiles as a defence against supply interruptions and developing alternative, particularly nuclear, sources of energy. And energy conservation policies often boiled down to using prices to depress fuel consumption, rather than regulatory measures such as fuel economy standards or buildings standards, or fiscal support for conservation. The rich countries remained split over prices: high prices suited the USA, which had oil producing interests, but not most European powers. Robert Lieber, an energy specialist who advised the US government, decried the lack of concerted action. Each country 'sought to insure itself against the worst', he complained; having been in 'disarray' after the first oil shock, the western powers again adopted an every-man-for-himself approach in 1979; West Germany and Japan stockpiled oil, further tightening the market and causing 'rationing by price'.[16]

The most significant supply-side measures focused on nuclear power stations. In the USA, there was a spectacular surge in orders for nuclear plants in 1973–4. But with interest rates running at record levels, the capital investment required brought several electricity companies close to bankruptcy. The last order was placed in 1978. The nuclear accident in 1979 at Three Mile Island, Pennsylvania, militated against the investment programme. All plants ordered after 1974 were cancelled, including some that were nearly completed. Nuclear electricity generation expanded more successfully in France and Japan, which both used fast-breeder reactors fuelled by plutonium.[17] The other supply-side policy departure was limited state support, in the USA and some European countries, for solar power and other renewable energy sources. But when oil prices fell in the mid-1980s, political commitment to renewables evaporated, and they remained marginal until the 2010s. (See pp. 34–5.)

After 1973, governments began to acknowledge that energy conservation, as well as alternative supplies, was a valid aim. But they almost always preferred to leave higher prices to trigger conservation than to invest in energy efficiency. In West Germany, a conservation programme proposed by the economy ministry in 1974 was rejected as detrimental to economic growth; proposals to reduce speed limits and to require higher fuel efficiency standards were blocked at the behest of car manufacturers and drivers' organisations. In France, too, conservation proposals drafted for government in 1977–8 were rejected. In both these countries, investments in energy conservation were small fractions of subsidies to coal production.[18]

In the USA, president Richard Nixon responded to the 1973 oil price shock by announcing a 'Project Independence' that comprised more nuclear power, more fossil fuel production, and some conservation measures. The political elite was split over the level at which to set fuel prices (i.e. to what extent the state would shield industrial and household consumers from the oil shock), and taxes and price controls on crude oil (i.e. to what extent the state would support domestic oil production). Since 1971, as part of an anti-inflation programme, controls had been put on prices of petrol, natural gas, and some domestically produced oil. In 1973, after the oil price shock, rapid inflation, queues at petrol stations and strikes and blockades by truckers dissuaded Nixon from even limited petrol price decontrol. In 1974, Nixon was succeeded by Gerald Ford; he proposed to lift the controls, but was beaten back by Congress.[19]

The extent of US political will to cut fuel use was again illustrated by the Energy Policy and Conservation Act 1975. On one hand, the Act required car manufacturers to meet fuel efficiency standards (the Corporate Average Fuel Economy standards) for the first time, and introduced fines for non-compliance. On the other hand, the rules were full of get-out clauses: manufacturers had only to achieve fleet averages, and so were free to build more fuel-inefficient cars if they were balanced by other models. And the standards were not onerous: with a 14 miles per gallon (mpg) average in 1975, they required that by 1978 new cars do 18 mpg, and by 1985 27.5 mpg. The carmakers were split: GM and Chrysler lobbied furiously against the new law, while Ford and other smaller companies accepted it as a lesser evil. For all of them, the regulations were a threat less serious than competition from Japanese carmakers that produced cheaper, more efficient models, and than the effects of the recession.[20]

What of more wide-ranging energy conservation policies? Certainly, the mood had warmed to them. Even before 1973, environmentalism had entered the political mainstream. In the USA, 20 million people participated in the first Earth Day in April 1970; Nixon then set up the federal Environmental Protection Agency (EPA) and signed the Clean Air Act, which regulated air-polluting emissions. In 1972, the UN Conference on Human Environment in Stockholm and the publication of the Club of Rome report *The Limits to Growth* made sustainability – as it came to be called later – part of international political dialogue.[21] (See pp. 173–4.) After the oil price shock, and amid the talk of an 'energy crisis', those who questioned society's dependence on fossil fuels were given a hearing.

In 1976, the energy conservation advocate Amory Lovins presented in US Congress the argument for 'soft energy paths', policies to raise energy efficiency, deploy 'soft' (resource-light) technologies, and to begin a transition away from fossil fuels. He denounced the 'bizarre notion that not to use more energy [...] means somehow a loss of prosperity'. Pointing to Denmark he argued: 'a rational energy system can virtually eliminate conversion and distribution losses that rob us of delivered end-use energy'. 'Technical fixes' that are 'now economic by orthodox criteria', e.g. thermal insulation, more efficient cars and cogeneration of heat and electricity, could slash consumption without any significant effect on lifestyles. Lovins, who was ridiculed by opponents for advocating 'a new dark age', continued until the present to argue that energy conservation strategies are compatible with a capitalist economy and US perceptions of prosperity.[22]

Lovins's far-sighted proposals were seriously debated by politicians, but US energy policy remained focused mainly on supply-side issues. In 1977, when Jimmy Carter succeeded Ford as president, he attempted a more coherent energy strategy, and proposed to Congress more stringent fuel economy standards, mandated efficiency standards for industry and tax credits for conservation measures. The package became law as a set of watered-down half measures. Perhaps the most significant change Carter made was the Public Utility Regulatory Policies Act (PURPA) of 1978, which challenged the dominance of the electricity market by big utility companies that owned fossil-fuel-fired and nuclear power stations. PURPA opened some space for independent power generators, for industrial firms that cogenerated electricity with heat to sell spare supplies into the grid, and even for wind-powered generation. As for petrol price deregulation, Carter pushed it through only in early 1981, after the second oil shock.[23]

Internationally, some energy specialists advocated a strategy that allied energy conservation with measures to confront the yawning inequalities in global energy systems. Living standards outside the rich world could rise substantially with support from an energy system that provided 1kw of useful energy per person, they argued. To conventional assumptions that energy supply had to rise to ensure economic development, they counterposed a focus on useful energy and on conservation through technological systems.[24] Separately, the 'appropriate technology' movement, that advocated small-scale, decentralised systems, particularly outside the rich world, briefly gained a hearing in the UN and development agencies – but by the mid-1980s had been left behind by policies focused on large-scale electrification.[25]

The new interest in energy conservation and its limits was reflected in academic research. Funding became available for net energy analysis. (See pp. 203–4.) In 1974, *Energy Policy* journal initiated discussion on the energy costs (as distinct from economic costs) of production processes. In 1976, William Nordhaus, the archetypal establishment economist and soon to become adviser to the US president, told an international gathering that the emphasis of energy research had shifted 'to the demand side of the equation', with special reference to conservation.[26] But most of the grand energy forecasting projects commissioned in the wake of the oil price shocks told a different story. The World Energy Conference, an international industry body – using methods that made no special study of sectoral demand or of useful energy – forecast in 1978–83 that energy demand would increase by 3–4 times between 1980 and 2020, and that most of the new supply would need to be met by nuclear power.[27] Studies by the US-government-linked Workshop on Alternative Energy Strategies (1977) and the International Institute for Applied Systems Analysis (1981) reached similar conclusions; the former warned, in particular, that oil supplies would be unable to meet demand past the 1990s. These studies assumed that robust economic growth should and would continue, and that energy demand growth rates would be driven by it. The dissenting voice was that of the World Resources Institute (1987), which argued that governments could intervene to delink energy demand from economic growth – a proposal that was largely ignored.[28]

Impact outside the rich world

Rich countries were made vulnerable to the oil price shocks by the proliferation of technological systems reliant on cheap fossil fuels; by contrast, developing countries that relied heavily on imported oil and oil products were mostly just starting to industrialise and electrify. The oil they used often went to vital infrastructure (commercial transport fleets and a few power stations) in economies otherwise dominated by biofuels. Oil products, mainly kerosene and diesel, fuelled lamps, water pumps and other essentials in the countryside. The heavier the import dependence, the tougher the choice: borrow to pay for oil imports, or abandon economic growth strategies. Hardest hit were small Central American countries that bought in more than 90 per cent of the oil they needed, The Philippines (more than 95 per cent), Pakistan, Bangladesh, South Korea (54 per cent) and Brazil (45 per cent).[29]

In some countries, development policies influenced by rich-country governments and the international financial institutions – and often reliant on technology bought from rich countries – exacerbated the crisis. Brazil was an example. During the post-war boom – and despite the fact that cars were for a small minority (one for every twelve Brazilians) – funds had poured in to build roads in the southern fifth of the country, where the urban population, and most of Brazil's motor vehicles, were based. Brazil became a hydrocarbons-based economy: oil and gas's share of total primary energy rose from 9 per cent in 1940 to 41 per cent in 1979. Some oil was produced domestically, but imports were needed. So when the oil crisis hit, Brazil spent much of the foreign exchange it earned from its own exports on oil imports – 47 per cent of it by 1982. The gravity of the situation made Brazil's military rulers receptive to unconventional energy policies, and, as well as allowing petrol and diesel prices to rise, they subsidised production of ethanol from sugar cane to substitute for imported oil. Ethanol output more than quadrupled, to more than 2.1 mtoe, between 1975 and 1981. Agriculture suffered for road transport. Land was removed from food production, and its ownership further concentrated, in a country where 40 million peasants were landless and cars affordable only for a small minority.[30]

Developing countries' oil import dependence, and the debts they ran up to pay for it, were measured by the economists Paul Hallwood and Stuart Sinclair. Sixty-one countries, with aggregate population of 485 million, relied on imports for more than 75 per cent of their commercial energy. (Of these, 32 were very poor countries that consumed less than 0.14 toe per head per year.) Another ten countries, with aggregate population of 313 million, relied on imports for 25–75 per cent of their commercial energy. Six countries, including India, with aggregate population of 709 million, relied on imports for less than 25 per cent of their commercial energy. The total oil import bill for the 87 largest non-oil-producing developing countries ballooned to $21.5 billion in 1974. In 1972 only Brazil had an oil import bill that comprised more than 10 per cent of its total import bill; in 1974, 21 countries did. The aggregate debts of these 87 nations grew from $120 billion in 1974 to $211 billion in 1977. In sub-Saharan Africa, oil importers often not only went into debt, but started Structural Adjustment Plans linked to loans from the International Monetary Fund (IMF), which imposed further hardships. They devalued their currencies, which hit living standards, and experienced food supply crises.[31]

Financially, the heaviest burden fell on countries that exported non-energy commodities. As recessions bit in the west, the prices of metals, minerals and agricultural commodities continued to rise – but much more slowly than oil prices. Attempts to form cartels to support prices were usually unsuccessful; and increasing export and other taxes on foreign businesses that dominated the commodity industries met stiff resistance. Jamaica, for example, tried to raise export taxes on US-owned bauxite producers; they responded by cutting investment. Brazil imposed aggressive export taxes on coffee in the mid 1970s, but these had to be withdrawn in 1978–9 as international prices tumbled.[32]

Governments' hands were forced; development strategies were dropped; electrification and industrial development plans were scrapped or postponed. Governments that borrowed to pay for oil imports found themselves unable to service debts to suppliers, and power station construction projects suffered. The cost of fuel subsidies soared – whether, to take India as an example, they were for LPG, to encourage poor households to move away from biofuels for cooking, or for widely criticised cheap electricity provision to farmers. (See pp. 115–16.)[33]

The oil price shocks had a devastating impact at the edges of the developing countries' islands of commercial energy supply. Rising kerosene and LPG prices often forced people who used those fuels for cooking or in cottage industries to revert to biomass. This in turn exacerbated the pressure on supplies of these non-commercial fuels, in the first place for traders who supplied wood to the urban poor, rather than for rural families who relied less on felled trees than on brushwood and twigs.[34] Deforestation became widespread. Rural families living near deforested areas then had to mount ever-longer expeditions to find wood for their daily needs.

In India, where biomass still comprised 90 per cent of household consumption, the poorest parts of the countryside felt a perverse knock-on effect of the oil price shock. In the mid-1970s, plans had been laid for large-scale tree planting, to replenish over-farmed forests. But with oil imports costing $972 million more in 1974 than in 1972, and the tripling of grain and fertiliser prices, development funds were diverted elsewhere. In many villages, people resorted to using cakes of dried cow dung as fuel, robbing farmland of badly needed nutrients and damaging soil quality. The government's agricultural commission declared that this use of dung was 'virtually a crime' – but desperation, borne of poverty, spoke louder.[35] Development researchers feared that the fuel wood crisis would spiral out of control, but fuel wood consumption growth slowed in the 1980s, and chronic exhaustion of resources was avoided.[36]

China's development story struck a contrast with India's. In the 1960s China had become self-sufficient in oil, and actively developed its coal and hydro resources. By the end of the 1970s, China was already the world's fourth largest producer of commercial energy (after the USA, the Soviet Union and Saudi Arabia), and the third largest consumer (after the USA and the Soviet Union) – but remained a predominantly rural and agricultural economy. In the 1970s there were repeated, acute shortages of basic household energy supplies, especially firewood, in the countryside. One set of policy responses used small-scale technologies – local coal mines and dams, and biodigesters – and had mixed results, for reasons that remain contested. It was centralised electricity networks that served rural industry and large farms before others, which finally began to provide an alternative to the biomass on which the countryside had relied for centuries. (See pp. 113–14.) This biomass had supplied about 90 per cent of rural energy consumption in the early 1970s; this would fall below 50 per cent by 1988.[37]

Developing countries that exported oil, such as Venezuela and Nigeria, also fared differently after the oil shocks. In 1974–8, the Nigerian government earned a $40 billion windfall via the state-controlled oil company, and in 1975 adopted an ambitious development plan including investment in oil and gas production, the development of a steel industry and transport infrastructure. Not only did many of these projects fail to materialise (the Ajakaouta steel complex never produced any steel, for example), but the country's own energy system remained into the twenty-first century overwhelmingly dominated by biofuels.[38]

Both in the rich world, and outside it, the oil price shocks and economic crises of the 1970s were a turning point. The boom-time combination of strong economic growth and cheap fossil fuels was finished. In the next chapter, the chronology is interrupted and patterns of electrification over the long term surveyed; in Chapter 8, the story of fuel consumption trends is continued, with an account of the 1980s.

7
Patterns of electrification

Electrification is central to the story of fossil fuel consumption. In 1950, one tenth of the world's fossil fuels were used to produce electricity; by 2011 that had risen to more than one third. In 1950, access to electricity was patchy in the rich world, and available only to small, urban islands outside it. In 1970, most (51 per cent) of the world's 3.7 billion population, including three quarters of people outside the rich world, had no electricity. Two decades later, in 1990, about 3.2 billion of the world's 5.3 billion population had electricity, including just under half of those living outside the rich world. Another two decades later, in 2013, 6 billion of the world's 7.2 billion people had access, including more than three-quarters of people outside the rich world. (See Table 3, p. 42.) 'Access to electricity' does not mean secure or adequate access: in addition to the 1.2 billion people who did not have it at all in 2013, a larger number had supply that was so limited and/or erratic that they could not depend on it. Moreover, governments who saw electrification as the key to economic development almost always prioritised supply to industry and agriculture, not household provision. Nevertheless, the number of people with some sort of access to electricity more than tripled between 1970 and 2013, while the number without it fell by almost one third.

In this chapter, some international trends in will be identified, and illustrated by five examples: the Soviet Union, China, India, Nigeria and South Africa.

International trends

During the post-war boom, electrification became a political priority in a large number of developing countries, and in much of the Soviet bloc. Non-OECD electricity generation grew from 130 billion kWh in 1950 to 2900 billion kWh in 1980. Electrification was usually undertaken by state-owned utilities, as part of government-led initiatives to develop infrastructure and industry. The electrical equipment industry, dominated by GE, Westinghouse, Siemens-AEG and a handful of other US, Japanese and European firms, acted as a driving-force of expansion. So did the newly created World Bank, which supported the state-led development model, and preferred large projects to less complex and smaller-scale technologies supported by some development specialists. By the early 1980s, World Bank loans for grid development totalled $17.8 billion, the bank's third largest portfolio after agriculture and transport; the largest recipients were India, Brazil and Colombia.[1]

In the aftermath of the Second World War, developing countries typically had a few small power stations, using diesel or small steam engines, to supply local networks in rich areas. From the 1950s, Walt Patterson wrote,

> [T]he drive for industrialisation prompted a parallel drive to install central-station electricity on a large scale. In the rush to electrify, city neighbourhoods were not top priority. As had been the case in industrial countries decades earlier, the initial focus was on providing electricity for industry. The technology of choice wherever it was available was hydro, feeding a synchronised AC network.[2]

South Africa, where electricity was produced mainly from coal rather than hydro, was an extreme example of this. Transmission lines ran right past black residential areas to supply white-owned mines and factories. But the priorities were similar everywhere: electrification strategies privileged industry over population, and cities over rural areas, both in capitalist countries, and in the Soviet Union and China.

Rural electrification is more capital intensive, and therefore more expensive, than providing access for city dwellers. It was sometimes pushed forward to support agricultural producers, but rural households, almost always and everywhere, were at the end of the queue for access to centralised systems. Of the billions of dollars invested in electrification globally between the end of the Second World War and the mid-1980s, an estimated 10 per cent went on rural systems.[3] The World Bank did not lend to projects specifically directed at rural electrification before the mid-1970s, although some rural areas may have been hooked up to transmission lines carrying current elsewhere; in the late 1970s the Bank was devoting just 8 per cent of its electricity project lending to rural areas.[4]

In the 1970s most of the world's rural population had no electricity – including 96 per cent of Africa's, 85 per cent of Asia's and 77 per cent of Latin America's, according to the World Bank. Even these estimates were probably overstated, since authorities frequently counted a village as electrified when only a small number of homes in it actually had an electric cable. The rural consumers that did have electricity used very little – from 200 kWh per capita per year in Thailand to 1000 kWh per capita per year in India – and most of that went to agricultural and commercial activities.[5]

Globally, rural electrification accelerated from the 1980s; since then, it has moved forward more rapidly than population growth. In the 1980s, some electricity systems in developing countries – caught between rising demand from expanding economies and growing populations, shortages of capital exacerbated by the 1970s oil shock, other economic pressures, and often conflicting requirements of government policies – found themselves in financial difficulties. The answer proposed by the international financial institutions, and embraced by some governments, was to open up the electricity sector to market reforms. (See pp. 143–7.) With respect to electrification, three points need to be emphasised. First, the 1990s reforms hardly ever followed the textbook model proposed, of privatising electricity; more often, they produced hybrid (part private, part state-owned) systems. Second, the progress of electrification in the 1990s mainly resulted *not* from privatisation, but from state and state-directed investment, in China and India for example. Private investment, if it produced any progress, did so mainly by improving existing networks. Third, in the poorest countries with the greatest need for electrification, market reforms in many cases slowed it down.

This third point was illustrated by a comparative study of 15 developing countries, published in 2004, which found that 'market-oriented reforms have either had a neutral or adverse impact on the poor'. (This conclusion was echoed by a clutch of other reports.) Key negative impacts were: reduction in electrification rates; increased tariff levels; and lower electricity consumption. Countries where reforms improved electrification – the Philippines, Thailand and Vietnam – had in common 'a high level of government involvement and special focus on protecting the interests of the poor'. In most cases, though, market-led reforms were 'primarily designed to improve the financial health of electricity companies'. Rather than expanding electricity access for poor people with unstable incomes, some of whom live in isolated areas, the companies unsurprisingly tended to '"cherry pick" the most lucrative markets', raise tariffs, and ignore the need to expand networks. In Kenya and Senegal, rural electrification rates fell in the immediate post-reform period; in Mali, they stagnated. In these countries, and Uganda, concerted efforts to address the pitifully low level of electrification began only after the reform had been completed, not as part of it.[6]

Electrification has often been constrained not only by the cost and difficulties of installing physical infrastructure, but also by the difficulties consumers – especially households outside the rich world – have in paying for supply. Tens of millions of poor households lacked, and lack, the funds to make the initial investment in a connection worthwhile; tens of millions more suffer such poor-quality service that they are unable to benefit from access, and end up off the grid again.

In 1965, a survey of rural India showed that once a village was connected to the grid, electricity connections spread to about 20 per cent of homes, but then demand growth slowed 'drastically'. A 1982 field study estimated that only 10–15 per cent of homes in electrified villages could afford a connection. In Bangladesh in 1978, 15 years after electricity had reached the Comilla area, less than 5 per cent of the rural population used it. The urban poor used electricity for lamps, but kerosene for cooking; they would acquire radios, fans and irons, and then refrigerators – but were always constrained by their ability to pay. A study published in 1979 showed that 'even where subsidised, electricity has not generally proven a suitably inexpensive energy source for household consumption by the poor', since the cost of the initial connection was often prohibitive. Although very poor people benefited from electricity even if they did not have it at home – because, for example, local shops and health clinics had refrigerators, and streets were lit at night – electricity as a consumer good, 'unless heavily subsidised, is likely to be limited to a small group of the relatively well off'.[7]

In the twenty-first century, households' inability to pay remained a major constraint on electrification. In many places, poor families in towns and countryside alike were connected, failed to pay, and were disconnected again. 'Even in villages that have been connected for 15–20 years, it is not uncommon for 20–25 per cent of households to remain unconnected,' a World Bank survey concluded in 2008. 'The absence of credit markets means households can not borrow to pay the connection charge.' Electricity consumption by poor families in the countryside in many countries remained at a very low level, a report published in 2010 showed: in Peru, 30–50 kWh per month per household, and usually less than 20 kWh in isolated areas, compared to 50–100 kWh in cities. In Thailand, five years after villages were connected, households had increased their consumption only from 11–22 kWh/month to 22–50 kWh/month, presumably because of their inability to pay for more. Cost was the main problem for the urban poor too.[8] Poor quality of supply also persisted. A 2008 survey of Indian households recorded an *average* of more than five hours per day of power cuts. Research of (supposedly) electrified households in Obantoko, Nigeria, in 2013 showed that no household had electricity access for more than six hours per day, and more than one-third had access for less than an hour per day. Nigeria's estimated national average was 16 hours per day of access.[9]

The following sections summarise some key features of electrification in five countries.

The Soviet Union

The government of Soviet Russia (from 1922, the Soviet Union) was the first to undertake state-directed electrification of a predominantly rural country outside the rich world. At the time of the 1917 revolution that brought the Soviet government to power, Russia had about 250 electricity utilities, most run on a concession basis and almost all reliant on small local power stations. The country only had 12 power stations with capacity greater than 5 MW, compared to 162 in the USA and 103 in Germany. Electricity had pride of place in the Communist Party's vision of modernisation, expressed in the slogan 'communism = soviet power + electrification'. In 1920, a powerful state agency, Goelro, was set up, dominated by engineers, socialists among them, who had long been convinced of the need for a centralised grid, and saw the Communists' accession to power as a golden opportunity to create one.[10]

The motivation of successive Soviet governments, and the elites they represented, was to ensure that industry in the former Russian empire could compete with that of the western powers. Electricity was a necessary condition of economic expansion; that, in turn, was seen as a necessary condition of the Soviet Union's survival and development. Whatever the slogans said, providing electricity for the population, particularly in rural areas, was a second- or even third-rate consideration. The same may be said of post-revolutionary China.

The 1920 electrification plan was launched amid an important debate around two contrasting approaches. The strategy adopted was to apply available resources to industrially oriented power stations, linked by transmission lines radiating through the most economically developed regions. Small-scale generation elsewhere, including almost all rural areas, would be left to local initiative. The second approach, supported by a substantial group of engineers but rejected on grounds of limited resources, was more actively to support a largely decentralised network of smaller power stations, enabling electricity to be brought to rural areas over shorter time scales.

At a conference of engineers in 1921, a representative from Kostroma, northeast of Moscow, said that a plan that electrified 'narrow belts' across the country, and left the rest of the population seeing electrification as a 'dream for the future', was insufficient. 'Local initiative' should be mobilised to stimulate construction of small- and medium-sized power stations; the national plan should integrate these. Advocates of a more centralised approach posed the dilemma in class terms: the peasant economy had to be subordinate to industry. Gleb Krzhizhanovskii, a leading Bolshevik, concluded that, ultimately, there were no resources to stimulate local stations. The energy historian Jonathan Coopersmith argued that more rapid rural electrification is one of the great 'what ifs' of early Soviet industrial policy. 'What if the party had promoted a voluntary, cooperative-oriented collectivisation, based on small power stations, in the mid 1920s, instead of the violent collectivisation of [1928–32]?' The centralised approach was 'technological determinism writ large', which cut across alternative possibilities for modernising the countryside, he proposed.[11]

The 1920 electrification plan envisaged building 100 new power stations. It was soon scaled back, as the economy was still recovering from the privations of the First World War and civil war. Until the mid-1920s, electrification hardly reached further than Moscow, St Petersburg and the oil city of Baku in Azerbaijan. Thereafter, construction of regional stations – usually coal, peat or hydro, depending on local conditions – got going, and, with forced industrialisation and collectivisation from 1928, accelerated. A target of installing 1750 MW nationally by 1932 was surpassed, mainly by adding capacity to existing stations. The forced march continued through the 1930s, reliant to an increasing degree on prison labour, which was deployed to build hydro stations on the Volga river, at Solikamsk in the northern Urals and on the great eastern Siberian rivers. In 1940 the security police accounted for nearly a quarter of total estimated investment in power station construction.[12] In that year, electricity output was an estimated 48 billion kWh, more than 20 times its 1913 level. US researchers estimated the electricity output growth rate for the USSR in 1925–50 at 10.5 per cent per year, the world's fastest.[13]

Soviet planners had a remit to bring electric lighting and appliances to farms where possible, but in practice that meant only those near the transmission lines supplying industry. By the eve of the Second World War, three large centralised systems, in the Central, Southern and Urals regions, were in operation, together with a smaller one in the Leningrad region – but only 4 per cent of state farms had any electricity. These were the foundations of a dual electricity system that emerged after the war. The electrical power ministry, which funded and managed large central stations and high-voltage links, was until 1953 prohibited by law from supplying state farms, and some other categories of consumers. The farms relied instead on a galaxy of unconnected local stations, of which there were 150,000 in Russia and Ukraine. The most basic suppliers were thermal stations serving agriculture: in 1967 their average capacity was 54 kW, and they generated electricity on average 2–3 hours per day. In 1965 there were 212,000 of these, more than 98 per cent of which had capacities below 500 kW. The lowest of the low were villages with no electricity at all. Abel Aganbegyan, an architect of Soviet economic reform in the 1980s, recalled that in the 1950s he had visited his wife's family in a village 120 km from Moscow that had no electricity, no rail access and no shops.[14]

The breakthrough in rural electrification came in the late 1960s, when means became available and political priorities shifted. By 1976, 99 per cent of rural electricity consumption was from unified networks. Milking and watering operations on farms were overwhelmingly electrified, although cleaning and feeding operations lagged behind. But industry still remained the prime customer for Soviet electricity, accounting for two thirds of consumption in the mid 1970s. Between 1928 and 1970 the growth in labour productivity was found to be closely correlated with the level of electric power generated per worker.[15]

China

China's electrification went together with successive stages of industrialisation that followed the 1949 revolution. More than 400 million Chinese people, mostly in the countryside, lived without electricity in 1949; that number fell to 245 million in 1979, when economic reforms began, to 20 million in 2004, and to a mere 1 million in 2015. During that time the population more than doubled in size, from 554 million in 1950 to 1407 million in 2015. In other words, electricity access expanded faster than the population did.[16] This does not mean that electricity became a ubiquitous energy source for most people, as it did in the rich world. Nationally, in the early 2010s, only about one fifth of final energy consumed was delivered as electricity, and most of that to industry. In the countryside only one tenth of energy was delivered as electricity, compared to four-tenths from coal and three-tenths from biomass.[17]

Electrification took place in three stages: (i) 1949–77 (state-directed economic development); (ii) 1978–97 (market reforms); (iii) 1997 onwards (rapid economic growth). In all these periods, electrification was financed overwhelmingly by the state. And at every stage – notwithstanding substantial investment in hydro power development, and, more recently, in gas-fired power – more than four-fifths of China's electricity was generated from coal.[18]

The Communist Party brought to power in 1949 inherited an economy ruined by a quarter of a century of almost continuous military conflict, and a total installed electricity capacity of just 1800 MW (equal to the capacity used by twenty-first century Cyprus). By 1957, capacity had nearly tripled, thanks largely to the supply of equipment from the Soviet Union, Czechoslovakia and East Germany. During the 'great leap forward' of 1958–9 – a generally disastrous attempt to accelerate industrial development that resulted in economic slump and widespread famine – China nevertheless registered continued expansion of generation capacity, although the level of electricity generated slumped in 1961, along with the whole economy. Then began a period of state-directed recovery: as in the Soviet Union, resources for electrification were devoted almost entirely to large-scale electricity projects for urban areas and industry.[19]

In, the countryside, where the vast majority of Chinese people lived, electrification was largely a spillover of industrialisation policy. Rural areas close to railways, roads and mining projects were the first to benefit. As local authorities endeavoured to raise agricultural production, they installed electricity for irrigation and drainage schemes. Central government policy, in contrast to the Soviet Union's, encouraged local authorities to develop autonomous electricity networks. In 1963 a government directive specified that national and provincial networks would provide electricity for irrigation in grain production areas; other rural consumers would be served by local systems, using small coal-fired stations and small hydro projects. The success of these policies was striking. In 1957, the countryside consumed less than 1 per cent of all electricity generated. By 1978, the countryside's share of consumption had risen to 13 per cent, and electricity networks reached 87 per cent of rural townships, 61 per cent of villages and 53 per cent of households.[20]

By the end of the 1970s, China's rapidly expanding economy showed signs of being short of electricity. There were staggered electricity cuts for industry, transmission load problems and a nationwide conservation campaign. In the countryside, many households had sufficient electricity for lighting only, and over-reliance on firewood for cooking caused deforestation in some areas. In the period 1978–97, the government addressed these problems in the context of market reforms across the economy. In contrast to other developing countries, which invited rich-country firms to establish independent power producers (IPPs), China sourced capital overwhelmingly from central and local government. Local authorities were given more autonomy to raise funds for electrification, under a policy of 'he who invests also owns and operates'. Market pricing was used to facilitate investment planning. In the countryside, town and village enterprises became predominant in many economic sectors including energy. An estimated $134 billion was invested in electricity networks in the 20 years from 1981. Electrification had in the previous period almost exclusively served industry and agriculture, but now reached tens of millions of additional households.[21]

In the third period, from 1997 onwards, China's coal-fired electricity generation again expanded exponentially, feeding the export-led industrial boom that made China the world's largest fossil fuel consumer. (See pp. 159–60.) In the countryside, a significant change was the merging of national and local electricity networks. At the turn of the century, 28 per cent of Chinese counties were still served by autonomous grids without a connection to the national network; 42 per cent by autonomous grids with a connection to the national network; and 30 per cent by extensions of the national grid. The state electricity company, one of the world's largest, put resources into unifying the networks.[22]

India

When India became independent from British imperial rule in 1947, it had just 1713 MW of generating capacity (a little less than China had). In the countryside, less than half a per cent of its villages – and a smaller proportion of households – had electricity. Electrification figured prominently in discussions about how to modernise the economy. The Soviet Union's centralised investment allocation system and the US New Deal were both referred to in the parliamentary debate on the Electricity (Supply) Act of 1948. Mahatma Gandhi, in keeping with his belief in community self-sufficiency, advocated decentralised electricity supply, but Jawaharlal Nehru's vision of state-driven industrialisation, influenced by the Soviet experience, prevailed. The Act provided for a publicly owned federal system, managed by State Electricity Boards (SEBs) owned by India's state governments.[23]

Electrification in India, by contrast with China, never caught up with rising population. The 1971 census counted a population of 548 million Indians, of which fewer than 100 million had electricity access. In 1991, the census counted 846 million Indians, of which around one third (270–80 million) had electricity access – compared to more than half of Chinese people at that time. At the turn of the century, India could claim to supply electricity to more than half the population, having increased the number of people with electricity access from 171 million in 1981 to 806 million in 2000 – and by that time more than 85 per cent of China's population had electricity access. In 2013, 237 million Indians, out of a total population of around 1250 million, had no electricity, according to IEA estimates – compared to just 1 million Chinese people. (The figures are not that accurate, mainly because authorities do not take sufficient account of differing access rates within villages.[24])

India's first state-directed electrification efforts were aimed at industry, which in 1960 used 74 per cent of the meagre 16.7 billion kWh generated. In the 1960s and 1970s, demand from agriculture – overwhelmingly, to run pumps for irrigation – rose steeply. Electrical pump-sets, which replaced earlier diesel pumps and could reach deeper into the water table, drove electrification across India as a whole. During the 'green revolution' aimed at raising agricultural productivity, both government and international organisations encouraged their use. Between 1970 and 1999, their numbers increased tenfold; agronomists feared that non-renewable water sources were being mined unsustainably. There was striking differentiation between, and even within, provinces, in the terms on which electricity was provided: where farmers were richer and politically well organised, as in Punjab and Maharashtra, cheap or even free electricity for irrigation became a mainstay of agricultural policy. Researchers found that corrupt provision of cheap electricity, and outright theft, was widespread. By the end of the 1980s, agriculture's share of electricity consumption had risen to 25 per cent, while industry's had fallen to about 40 per cent (a figure that does not include the substantial amount of autogeneration, i.e. firms producing their own electricity, off grid).[25]

In 1991, the Indian government initiated electricity market reforms. Initially, the measures brought some foreign investment into generation, but left much else unchanged. (See pp. 144–5.) Electrification – which, by the 1990s, meant mainly improving access for poor households in the countryside – was 'not recognised explicitly as an objective' of the reform, in the energy researcher Shonali Pachauri's judgement. Rural access, and other issues of electricity distribution, did not begin to be addressed until the end of that decade – against resistance from the electricity utilities, who continued to insist that rural electrification was unviable.[26] The reforms deepened the fault lines that had run through electrification policy since the 1970s, and the differences between states, in the historian Sunila Kale's view. Her study highlights the following examples.

In *Maharashtra*, well-off farmers, doing energy-intensive sugarcane farming and cane crushing, were strongly represented in state politics. This ensured that they gained most from rural electrification in the 1970s. A system of flat-rate tariffs for electricity effectively pushed an outsized share of the state's total subsidies to these 'sugar barons'. Their political weight was sufficient to ensure that in the 1990s, when reforms supposedly aiming to reduce subsidies, theirs remained untouched. Political leaders in the state continued even into the 2000s to make election pledges of free electricity to farmers.[27] These conditions made Maharashtra less attractive to opportunistic international investors than other states; its largest IPP project, the Dabhol plant operated by Enron, provoked widespread protest at its high cost and lack of transparency, and the deal broke down.

In *Gujarat*, as in Maharashtra, farmers organised into a powerful political lobby in the 1980s to secure favourable flat-rate tariffs. By acceding to these demands, the state government found itself unable to regulate electricity supply effectively, in Navroz Dubash's view. The 1990s reforms were judged to be relatively successful in Gujarat: the State Electricity Board company was unbundled and scheduled supplies for irrigation introduced with a view to controlling flows. But disputes over tariff setting persisted past the turn of the century.[28]

By contrast to Maharashtra and Gujarat, *Odisha* (former Orissa), one of the poorest states, until the 1980s pushed the lion's share of its electrification resources towards industrial development and urban consumers. Where rural areas were electrified, the social and economic benefits were soon obvious, but most of the countryside remained without electricity access. When the reforms began, Odisha's urban generation assets, and its strongly pro-market state government, seemed attractive to investors. The World Bank and its consultants arrived in force, hoping to make the state an example for others to follow. Generation companies were allowed to increase the tariffs they charged to the state-owned distribution network, but its tariffs to final consumers were held down; 75 per cent of shared financial liabilities were transferred to the state. (See p. 145.) The reforms were termed a success, but in the immediate post-reform period, while the level of electrification of better-off rural households rose, the level among the poorest households fell, as families found themselves unable to pay higher charges.[29]

In *Andhra Pradesh*, there was no strong farmers' lobby analogous to that in Maharashtra. It took a movement of less well-off farmers in the 1980s – representing groups that, in Kale's view, had 'felt marginalised by successive Congress-dominated state governments', and agitated for inputs and price supports, including subsidised electricity for irrigation pumps – to push rural electrification to the fore in state politics. This movement, in turn, influenced events in the 1990s. Its participants, and others in rural communities, were resistant to privatisation, and it was unsuccessful.[30]

All these examples underline that social forces – farmers' lobbies such as in Maharashtra, the broader rural movement in Andhra Pradesh and local

Nigeria

Nigeria is an example of a country that until very recently had no electrification strategy worthy of the name. Electricity production has risen many times over, from a meagre 0.75 billion kWh per year in the mid 1960s, to 7.7 billion kWh in 1981, 15.4 billion kWh in 2001 and 25–8 billion kWh per year in the 2010s. The population – Africa's largest – grew fivefold between 1950 (when it was 37.8 million) and 2010 (159.7 million). In human development terms, electricity provision has been running to catch up, but never did so. By the 2000s, per capita electricity consumption was less than one-ninth of Egypt's and less than one-thirtieth of South Africa's.[31] Nigeria's emergence since the 1970s as sub-Saharan Africa's leading oil exporter not only failed to enhance energy provision to its citizens, but may have worsened it.

The electricity access rate was by the 2000s estimated at 40 per cent (78 per cent urban and 23 per cent rural), although the average connection was interrupted 60 per cent of the time. Nigeria's urban population and industry has long felt electricity to be in chronically short supply, making autogeneration a major industry. In 2013 Nigeria had an estimated 60 million generating appliances; 85 per cent of businesses had their own generator; and 80 per cent of households use non-grid sources of electricity including generators and solar inverters.[32]

Before Nigeria gained independence from the British Empire in 1960, electricity supply, like piped water, was 'limited, spatially and socially, to the Island [area of Lagos], and the elite (European and African) that lived there', the economic historian Ayodeji Olukoju recorded.[33] Only after recovery from the 1965–6 civil war, in 1972, did government create a central institution to manage electricity supply: the National Electric Power Authority (NEPA). It oversaw the state-owned Electricity Corporation of Nigeria and Niger Dam Authority, and private producers who supplied industry. Between 1964 and 1983 generation capacity grew by 11 per cent per year, but every new power station brought only partial, temporary relief from shortages and electricity cuts. The biggest consumers, industry and the oil companies relied increasingly on privately owned generating equipment. The grid hardly touched the countryside, or the whole north of Nigeria with the exception of the industrial parts of Kaduna and Kano states.[34]

In the 1990s, with urban populations swelled by migration from the countryside, and elements of civil society developing despite military rule, dysfunctional electricity supply took on the character of a national scandal. Between 1970 and 1987, Lagos's residential customer base expanded fivefold, and their electricity consumption elevenfold. But by the 1990s, 'entire neighbourhoods could be in complete darkness for months; the more fortunate wards or streets had to put up with "load shedding", by which power was rationed [...] at the whim of officials', Olokuju wrote. In his view, this was the 'culmination of years of neglect or of wrong policies', exacerbated by the disrepair of the distribution network, ubiquitous illegal connections and large-scale corruption at NEPA, now popularly referred to as 'Never Expect Power Always'. Sales of generators, candles and storm lanterns boomed.[35]

After military government ended in 1999, Nigeria began the electricity market reforms that had started elsewhere throughout the decade. Nigeria followed the common practice of inviting foreign investors to establish IPPs. First to arrive was Enron, which agreed to supply NEPA's Lagos grid from nine barge-mounted gas turbines. Before going bust in 2001, Enron sold its stake in the project to AES. NEPA tried to renegotiate a 13-year power purchase agreement that locked in unsustainably high prices. This failed, and the deal collapsed in a slew of legal actions. The reform ran into the sand; a new version was rolled out in 2005. A law was passed providing for NEPA to be unbundled, generating assets to be sold and capacity to be added by a National Integrated Power Project. Privatisation, and Nigeria's first serious attempt at state-directed investment in electricity generation, thus came side by side. Existing plant was sold off, but investors' commitments to add new capacity produced few results. Only one of 23 firms involved had 'done anything tangible' by 2008, in Akin Iwayemi's judgment. New plans were made to expand capacity with investment support from China.[36]

In the 1990s, the contrast between Nigeria's standing as a major oil producer, and the poor state of its own energy system, became the subject of civil conflict. The execution in 1995 of Ken Saro-Wiwa, an environmental activist who led protests at oil companies' operations, attracted international attention. After the turn of the century, some campaigners focused on natural gas flaring at oil fields.[37] They pointed out that if most of the flared gas was put into power stations of average efficiency, Nigeria could double (at least) its electricity output and reduce its contribution to global warming. The absence of investment in this obviously beneficial reduction of waste, over decades, is telling. Exxon and Mobil Producing Nigeria both said they hoped to make such investment, when chasing IPP contracts in 1999; neither did so. Shell, the IOC with the largest presence in Nigeria, began to reduce flaring only in 2002, and sold a very small amount of gas to electricity producers. The estimated volume of gas flared has fallen, from 14.6 bcm in 2011 to 7.6 bcm in 2015, but not significantly to the electricity system's benefit. So Nigerian oil production continues to contribute disproportionately to global warming, while the population of oil-producing areas endures negative impacts, such as oil spills that poison water supply and funding of criminal militia from illicit oil sales ... and, even now, has near-zero electricity access.[38]

South Africa

In South Africa, the continent's most industrialised and electrified country, the electricity grid took shape in the colonial period, to serve the gold, diamond and coal mines. In 1948 it was nationalised with a view to serving these industries, and the white middle class, more effectively. The black majority was denied electricity access not simply by the market's inequalities, but also by government policy, which discriminated against blacks. There was no systematic electrification of black households until the late 1980s, and it was not widespread until after the racist apartheid system collapsed in 1994. The proportion of households with electricity access rose from 40 per cent in that year to 73 per cent in 2006.[39] The issue of how electricity supplies, especially to the urban poor, would be paid for was not solved, though. Nowhere did views of electricity as a public service, and as a marketable good, clash more sharply.

The first phase of South Africa's electrification began in the 1880s with the diamond mines at Kimberley. Gold mines were by far the most important consumer, though: plentiful electricity was a prerequisite for working deep deposits and low-grade ore. The mining companies' privately-owned power stations by 1920 comprised 'one of the most sophisticated energy systems in the British empire', in Leonard Gentle's account, generating as much electricity as London, Birmingham and Sheffield combined. In the 1920s, two monopolistic privately owned power companies emerged: Escom, which provided electricity to the railways, on which the mines depended heavily, and the Victoria Falls and Transvaal Power company, which oversaw the grandiose Victoria Falls hydro project, and the generation of coal-fired steam power in Eastern Transvaal. Under the racist 'segregation' system that preceded apartheid, black workers in South Africa – on whose labour the mining complex depended – were deprived of electricity at home. (South Africa became independent of the British Empire from 1931; apartheid was introduced in 1948.) The Orlando power station, commissioned in 1943, supplied the mines, but not the township around it: pylons from it, dwarfing un-electrified shacks beneath, became symbolic.[40]

The generating companies failed to invest sufficiently to meet the gold mines' needs – no less than 59 per cent of all electricity produced in the late 1940s – and in 1948 Escom was nationalised. Large long-term loans were raised from the World Bank, and in the 1960s three power stations built on the coalfields, to supply the gold industry. In that decade Escom also invested in transmission, bringing a national grid into being by 1973. Household electrification now began in earnest – for whites. In Johannesburg, white suburbs had access rates close to 100 per cent by the mid 1970s, compared to 20 per cent in Soweto, the adjacent black township. In the countryside, electrification was targeted at white farmers. Excess power station capacity was mothballed, rather than being mobilised to electrify black households.[41]

The mines, and above all the gold mines, remained the prime consumer of South African electricity. In the mid-1990s, mining and mineral processing accounted for 40 per cent of electricity consumption, and the economy was estimated to be three times as energy-intensive as the USA's. As for households, the government's attitude to the majority, who had systematically been denied electricity in the past, began to change in the mid-1980s as the anti-apartheid protest movement gathered pace. An 'Electricity for All' programme, aimed at selling electricity from surplus capacity to black townships, was launched in 1985. It first produced noticeable results in 1991, when 25,000 dwellings were connected. Household electrification received political impetus as apartheid ended in 1994. An average of 300,000 homes per year were connected, for a decade from 1992. Government policy prescribed a system under which private business provided electricity, but in practice electrification was supported by the state and financed by (the slightly renamed) Eskom, largely through a cross-subsidy scheme. By 2002, 79 per cent of households in urban areas and 46 per cent in rural areas had electricity access.[42]

The constraints on electrification were economic. In the 1990s, the municipal electricity distributors that supplied households were charging them about twice the rate that bulk industrial customers paid. There were widespread protests and political conflicts over electricity prices. (See p. 149.) In 2001 the government introduced free provision of water, electricity and some other municipal services to poor households. Those not covered by the scheme often rationed their own electricity use, in order not to be cut off: more than half of households were estimated to consume no more than 50 kWh/month. At the turn of the century, 78 per cent of African households had a TV, 58 per cent a refrigerator and 9 per cent a washing machine, compared to 98 per cent, 98 per cent and 91 per cent respectively in white households. Low-income households which had such appliances often turned them off to keep bills low.[43]

Some conclusions

In all five countries discussed, the initial electrification of city centres and key industries was undertaken by private companies, from the late nineteenth century. But electrification above this level – its development, as reliable infrastructure for industry and agriculture, and/or as a public service – was undertaken by the state.

In the Soviet Union and China – and to some extent in India – government-financed electrification was seen, ideologically, as the work of the modernising socialist state. The state socialist governments, no less than those of capitalist countries, prioritised the electrification of industry and, to a lesser extent, large-scale agriculture. In contrast to the Soviet Union, China adopted a more flexible policy towards rural electrification by local authorities. This enabled the Chinese countryside to race ahead of India's, achieving electricity access rates well above 90 per cent by the turn of the century. The private sector everywhere proved inadequate to the task of electrification – in South Africa in the 1940s, leading to nationalisation, in Nigeria throughout its history, and in India following the 1990s reforms.

Social and class forces shaped electrification everywhere. The state socialist elites undertook the most impressive projects, electrifying industry – and in China's case agriculture – decades before wider public service provision. Nigeria, subjugated first by empire and then by neocolonial economic exploitation, proved unable to develop an electrification strategy, despite being such a significant exporter of fossil fuels to world markets. The South African elite, which needed the electricity primarily for mining and processing gold and other minerals, systematically deprived its black population of electricity access until the fall of apartheid in the 1990s. In India, the well organised lobby of wealthier farmers in Maharashtra and Gujarat ensured earlier electrification of agriculture; the broader rural movement in Andhra Pradesh influenced the outcome of market reforms; and in Odisha, where such a movement was absent, the local elite and international organisations undertook reforms in a manner that de-prioritised rural electrification.

8
The 1980s: recession and recovery

The second oil shock of 1979–80 helped to cause a deep recession in the largest capitalist economies in 1980–82. In the rich countries, oil use fell sharply, total commercial energy use less sharply. Then, from the mid-1980s, renewed economic expansion, driven largely by globalisation and the export of capital to developing countries, produced a new boom in fossil fuel use. In electricity, the expected nuclear dawn never arrived: coal and gas filled the gap. In the 1980s, political battles over energy conservation intensified. The oil price shocks had produced noticeable results in terms of conservation, particularly in rich-world industries – but these were overshadowed by the waste of opportunities to develop sustainable consumption more widely. Governments rejected the sort of investment strategies and regulation that could have changed more profoundly the ways that energy is consumed – and supported incumbent fossil fuels with subsidies. These two policy trends prepared the way for a much greater disaster at the end of the 1980s: governments' collective paralysis in the face of global warming.

From recession to recovery

Total commercial energy consumption fell during the global recession for three years running (1980–82), and in the rich countries, oil consumption continued to fall for a further three years. In the OECD, oil consumption only regained its 1979 level in 1996, while gas consumption recovered more quickly (by 1990) and coal consumption rose almost uninterruptedly. In the five years to 1990, total OECD fossil fuel consumption rose by 8.8 per cent, and continued to grow nearly as rapidly during the 'roaring nineties'. Total non-OECD fossil fuel consumption rose faster in the 1980s: by 18.3 per cent in the five years to 1985, and by 16.1 per cent in the five years to 1990. That rate slowed substantially in the early 1990s, reflecting, in large part, the economic slump in the former Soviet countries. (See Table 4, p. 55, and Table 8, p. 97.)

Fossil fuel consumption remained highly concentrated. In 1987, 90 per cent of coal was consumed by 15 countries (led by China, the USA, the Soviet Union, Poland, India and West Germany); 80 per cent of petroleum products by 28 countries (led by the USA, the Soviet Union, Japan, China and West Germany); and 91 per cent of natural gas by 20 countries (led by the Soviet Union, the

USA, the UK and Canada). And the rich minority consumed the most: OECD countries, with less than 15 per cent of the world's population, accounted for 62 per cent of commercial energy consumption in 1980, and still for 58 per cent in 2000 – figures that exclude energy used to produce goods that they imported.[1]

The early 1980s were a turning point for the world economy. The recession of 1971–3 had ended the longest ever continuous period of economic growth and rising profitability in the rich countries; the second oil shock and stagnation of 1980–82 compounded the crisis. Economists point to a crisis of profitability that underlay the end of the boom, and see both globalisation (meaning the restructuring of capital and work processes, and financial deregulation), and the adoption of neoliberal economic policies, as responses to it. A 'long period of expansion' was 'initiated by tax cuts and spending increases in the USA' in 1983–9, one standard account underlined – although economic growth rates, and investment and productivity growth levels, remained well below those of the post-war boom. The international structure of industry began to shift. Drastic cutbacks in rich countries (e.g. the UK, which lost a quarter of its manufacturing industry in 1980–84) were followed by a massive, long-term expansion of industrial capacity in some developing countries.[2] Between 1980 and 2005, the global export-weighted labour force quadrupled, David McNally pointed out. About half of this growth was in east Asia, where the number of industrial workers rose from 100 million to 900 million. There were great differences between parts of the developing world. China and other Asian manufacture-exporters maintained high rates of investment growth, but the rate of accumulation fell sharply elsewhere, particularly in countries that relied on exports of oil or other raw material commodities, whose prices fell sharply in the mid-1980s.[3]

The restructuring of capital and of work processes fundamentally changed social relations. In the rich countries, frameworks for compromise between capital and labour that had been in place during the boom broke down. Capital sought new means of social control. Part of the strategy for restoring profitability was confrontation with labour unions; a sustained attack on wage levels and other social benefits, underpinned by unemployment; and increases in labour productivity, using new technology and management techniques such as 'lean production'. Financial globalisation was also seen as a response to the crisis of profitability. The financial markets deregulation of the 1980s, enabled by new computer technology, helped to redirect oil dollars that had accumulated in the Middle East, Angus Maddison argued, and transformed the relationship between the USA and the rest. From being a net lender up to 1988, the world's pre-eminent economic power moved quickly to a net foreign asset position of minus $1.5 trillion (more than 20 per cent of its GDP); in this way, the rest of the world helped to sustain a renewed American boom and financed the US payments deficit.[4]

Energy markets were remade against this background. The price of oil plunged from $34/barrel in 1982 to $12/barrel in 1986. There was overproduction for the first time since 1928. The geography of oil supply changed. Supplies from non-OPEC producers, including the Soviet Union and the UK, grew: OPEC's share of world oil sales, which had been 50 per cent in 1973, fell below 30 per cent by 1985. The big western economies' dependence on OPEC oil supplies, 'absolute' in 1973, became merely 'acute and dangerous' by 1988 when the Iran-Iraq war began, Paul Stevens suggested. The way the market worked also changed. The largest domestic market, in the USA, was deregulated, and a world market dominated by the IOCs began to give way to a traded commodity market. In 1986, 30 per cent of oil was sold on a 'spot' basis (i.e. not covered by contracts agreed in advance), and that proportion grew steadily thereafter.[5]

In the early 1980s, economic slowdown and industrial restructuring had helped to push down oil demand. So did fuel switching, mainly to gas and coal. In the 1970s, western governments had seen natural gas as a cooking fuel and as raw material e.g. for premium aircraft fuels. Burning it to generate electricity was prohibited in the US (between 1978 and 1987) and constrained by regulation in the European Union (between 1975 and 1991). Attitudes, and regulation, changed in the 1980s. The invention of the combined-cycle gas turbine ramped up gas-powered electricity plants' efficiency. (See p. 29.) All this boosted gas demand. The Soviet Union, with West Germany and other European countries, in 1981–5 built the big pipeline corridor from the western Siberian gas fields to western Europe, notwithstanding the US government's opposition to expanding east-west trade.[6] The Soviet Union, and after its collapse Russia, became and remained Europe's largest supplier of imported gas.

Another reason for the steady growth of gas and coal supplies in the 1980s and 1990s was that the nuclear age expected by energy specialists and politicians alike never materialised. In the early 1980s, it was routinely forecast that nuclear power would become a leading, or *the* leading, source of incremental energy supplies by the turn of the century; fast-breeder reactors would prevail.[7] But the USA's Three Mile Island accident in 1979, and the reactor fire at Chernobyl, Ukraine, in 1986, stimulated opposition from social and environmentalist movements. Massive cost overruns made matters worse. The nuclear sector grew, especially in France and Japan, but never achieved the role imagined for it.

The next two sections cover fossil fuel consumption trends in both the 1980s and the 1990s.

Rich-world consumption, 1980–99

Across the rich world, the energy conservation gains produced by the two oil shocks were 'surprising even to conservation enthusiasts', one researcher wrote.[8] Substitution of capital and labour for more-expensive energy, and technological change, counted. There was a 'metamorphosis' in car design and production, the synthetic fuels business was 'stillborn', and architectural design 'visibly altered'. Overall, the OECD countries' energy intensity fell by 24 per cent between 1974 and 1986. Gone was the 'widespread assumption' of a 'God-given relationship' between the pace of energy consumption growth and economic growth, the energy economist James Jensen observed in 1985.[9] But after energy prices fell in the mid-1980s, many of the conservation gains were reversed.

In OECD countries, profligate consumption during the boom had left 'ample room for improving energy efficiency' when the 1970s price shocks hit – but in the mid-1980s, lower prices, combined with lower energy intensities, stimulated consumption and unwound what progress had been made, the IEA concluded. In 1982–6 energy use was constant but did not fall; in 1987–90, energy intensities declined more slowly and consumption started to rise again, at 2.2 per cent/year. The retreat from efficiency gains continued in the 1990s, in spite of international agreements on the need to tackle global warming. The 1970s oil price shocks, and resulting policies, 'had a larger impact on the increase in energy demand and reduction in CO_2 emissions than the energy efficiency and climate policies implemented in the 1990s', the Netherlands' environmental agency observed. Internationally, in manufacturing, energy intensity fell by 36 per cent in 1973–86, but only a further 4 per cent in 1986–98. The heart of the problem was in the USA. A UNDP study showed that US consumption fell between 1973 and 1983, as its population and economy expanded – but 'during the next 15 years, from 1983 to 1998, the USA lost all the gains in energy conservation it achieved. Americans returned to consuming nearly as much energy as before the oil shocks'.[10]

Modest reductions in the rate at which rich countries' economies used fossil fuels and other raw materials triggered speculation that they were 'dematerialising', i.e. that economic growth had decoupled from rising consumption of materials, and/or were using less materials overall. Researchers were already doubtful about these claims in the late 1990s. Economic growth and resource throughput had 'decoupled' in rich countries between 1975 and 1996 – but *overall* resource use and flows of waste materials had risen, a World Resources Institute study showed in 2000. Production of energy-intensive steel and aluminium had fallen in the rich countries, but the energy saving might be cancelled out by production of smaller quantities of more energy-intensive materials (e.g. plastics), and the effects of trade and higher consumption levels, the UNDP concluded in the same year.[11] Recent research has confirmed that *globally, overall*, in the four decades since the oil price shocks, materials consumption has risen consistently and 'decoupling' is limited and partly illusory. (See p. 155.)

Nevertheless, there *were* energy efficiency gains in response to the oil price shocks, most significantly in industry. In Japan, industrial energy consumption per unit of economic output had by 1985 fallen by 70 per cent of its 1971 level; the falls in the USA and western Europe were almost as dramatic. Most of the decline was 'attributable to conservation, restructuring and fuel-switching', James Jensen wrote. The export of energy-intensive industries to developing countries – in Japan's case, the removal of chunks of the petrochemical industry to Taiwan, South Korea and elsewhere – was another big factor.[12] In Europe, the energy intensity of industry fell between 1979 and 1986, firstly because steel and other energy-intensive sectors were severely cut back, while less energy-intensive sectors grew, and secondly thanks to efficiency gains.

The amount of fuel used to make steel was reduced, in rich countries, when companies deployed EAFs in mini-mills, in the context of cutting costs and labour forces. US Steel, the USA's largest steelmaker, brought in $2 billion worth of new equipment, including EAFs, and reduced the workforce from 93,000 in 1980 to 23,000 in 1991. Car factories, which accounted for about one-fifth of European steel use in the 1980s, cut the amount of steel per car – although not to anywhere near the technologically possible minimum – imported more of the steel, and increasingly used aluminium.[13]

From the mid-1980s, low energy prices reversed many conservation gains in industry. Casting aluminium more accurately, or recycling more efficiently, would have cut fuel inputs sharply, the energy researcher Daniel Spreng pointed out in 1988. 'Letting pumps, compressors, conveyor belts and so on run only when they are actually needed' would be 'potentially a big saving', he argued. In the 1990s, European industry raised its output by 2 per cent/year without increasing energy consumption. But, the UNDP argued, great potential remained for energy saving, especially in petrochemical plants, paper mills and parts of the steel industry, and for cogeneration of electricity and heat. Europe's constant energy demand struck a contrast with the USA, where industrial energy use fell by 18 per cent between 1973 and 1983, but grew by 37 per cent from 1983 to 1997.[14]

In the 1980s and 1990s in rich countries, the nature of work, and of personal consumption, was changed by three trends: the attack on wage levels and social benefits; transfer of industries to the Global South and growth of the service sector; and technological changes, especially those associated with computing. As a result, consumption became more fossil-fuel-intensive. The hours people worked tended to stabilise or rise, instead of falling as they had in the post-war boom. Sociologists observed a tendency for people to spend more time working, earn more, and, to make up for shortage of time, shift to more 'goods-intensive' lifestyles. They bought freezers, dishwashers and microwave ovens; more often ate food prepared in restaurants or take-away outlets; and used private, rather than public, transport. The economist John Maynard Keynes's 1931 forecast that his grandchildren's generation would have more leisure time was not borne out, researchers of consumption pointed out; instead, rich-world populations had been 'induced to be competitive', which drove them to keep 'working hard for more income to buy more goods'.[15]

The energy-intensive nature of such lifestyles was exacerbated by mass production techniques, combined with the outsourcing of manufacturing to developing countries. This put a vast quantity of consumer goods in reach of rich-world working families, and made it cheaper to replace goods than to mend them. The sociologist Juliet Schor pointed to the examples of TVs and personal computers, the prices of which collapsed, respectively, by 58 per cent and 86 per cent in the US between 1997 and 2005. Another bloated channel of fossil-fuel-intensive materials was plastic toys, mostly made in China; in the USA in 2001, $29.4 billion worth, or 69 per child, were sold, or (often) given away with fast-food meals.[16]

All these factors drove up direct and indirect energy use by rich-world households. Direct household energy use rose by 17 per cent between 1973 and 1998, an IEA survey of eleven rich countries recorded. And it accelerated towards the turn of the century: in 1994–2003 it was growing faster than ever. Space heating remained the most significant end use of energy in households, and while the amount of heat required per unit of floor area fell in most countries, 'bigger houses and fewer people per dwelling' cancelled out potential savings, the IEA found. Household coal consumption fell away, while electricity consumption doubled and gas use increased by 40 per cent. In the UK, household heating changed fundamentally for the first time in 300 years, with the appearance of pressure boilers and central heating systems, which were increasingly run on gas instead of coal. In 1970, 35 per cent of homes were centrally heated; by 1990 it was 80 per cent.[17]

Strong growth in ownership of household appliances propelled the increase in electricity consumption, the IEA found. Appliances became more energy hungry. In Europe, household energy consumption reached the saturation levels that the USA had attained during the post-war boom. In the Netherlands, penetration rates of TVs, refrigerators and washing machines rose from nearly zero after the Second World War to more than 100 per cent in 1990. In the UK between 1954 and 1994, the number of household appliances doubled, and the number of durable entertainment goods (e.g. TVs and stereos) multiplied 35 times, David Goldblatt found.[18]

A final significant cause of fossil fuel consumption growth across the rich world in the 1980s and 1990s was a renewed surge in private car use. Governments actively supported this boom, by keeping petrol prices down, reducing regulation of carmakers and prioritising road construction. The IEA reported:

> Real gasoline [i.e. petrol] prices peaked in the early 1980s and have generally fallen since then. In most European countries much of this decline was offset

by increasing fuel taxes. Still, real prices, including taxes, were lower in the late 1990s than in the early 1980s in almost all IEA countries.[19]

The US government was a significantly worse offender than others. Its own Energy Information Administration (EIA) calculated that the real price of petrol for end users fell by 45 per cent between 1981 and 1988, before rising gently in the 1990s. Overall, between 1973 and 1998, it rose by only 0.25 per cent per year. US federal taxes on petrol were raised in 1990, despite heated conservative opposition – but in 1991 real petrol prices in other rich countries were often 50 per cent and, in Japan and Italy, up to 300 per cent higher than the USA's.[20]

In 1985, the Reagan administration relaxed the already-feeble regulation of fuel efficiency standards. The standard for new cars then remained unchanged until 2007, at 27.5 mpg; the standard for light trucks was lowered in 1989 from 20.7 mpg to 20 mpg. These numbers were put in perspective by the physicist Frank von Hippel, who in 1981 showed that 60 mpg was technologically feasible.[21] US cars on average improved their efficiency between 1980 (19.2 miles per gallon) and 1988 (22.1 mpg), but fell back to 20.4 mpg by 2002. Vans, light trucks and sport utility vehicles (SUVs) improved, from 12.0 mpg in 1979 to 17.5 mpg in 2000 – but that amounted to a step backwards for each driver that switched from a car to a SUV. And there were millions of them. The big three carmakers broke SUV sales records year after year, culminating in the sale of 17.4 million SUVs in 2000. By then, more than half of the vehicles on US roads were classified as trucks. In Japan, too, vehicle efficiency stagnated and then worsened in the early 1990s, as vehicle sizes rose and petrol prices fell.[22]

In the USA, the total number of vehicle miles travelled rose by 82 per cent between 1980 and 2000. Traffic jams went from occasional to ubiquitous. Peak-hour road travellers in urban areas of Texas were delayed in traffic jams, on average, by 8.7 hours in the year 1982, and 39.2 hours in 2000, a more than fourfold increase. In large cities it was 76.5 hours in 2000. Part of the problem was the trend towards people driving alone: in 2000, the Nationwide Personal Transport survey recorded that 75 per cent of commuters travelled singly in privately-owned cars, and less than 5 per cent on public transport.[23]

Cars were slightly less pervasive in other rich countries. Even in the UK, which was more motorised than most, 30 per cent of households remained without a car up to the late 1990s – although cars had become cheaper to run, both absolutely and relative to using public transport. Between the 1970s and the 1990s, the proportion of women with driving licences almost doubled; new household tasks, such as taking children to school by car, were part of the cause.[24]

Consumption outside the rich world, 1980–99

Industrialisation, urbanisation, electrification and motorisation all contributed to the rapid increase of fossil fuel consumption in developing countries in the

1980s and 1990s. In many countries (but not China), manufacturing became more energy efficient after the oil shocks. But because fossil-fuelled systems expanded more generally, the level of primary energy use per unit of economic output rose (except in China). Motorisation, electrification and household consumption played their part. Prices of cars and household appliances had fallen in real terms since the 1970s, and now they were often made locally, UNDP researchers found. In households in higher-income developing countries, cookers and water heaters had become more efficient, but their rapid diffusion far outpaced the efficiency gains.[25]

A blizzard of motorisation hit developing-world cities, encouraged by international agencies and governments who ignored everything urban planners had learned about the damage done by car-based cities in the rich world. In the 1980s, the Nigerian capital, Lagos, began to experience its mind-boggling traffic jams, thanks in large part to unplanned city expansion. By 1990 Mexico City had an estimated 4 million cars, 40 times as many as in 1950. When the Soviet bloc collapsed in 1989–91, levels of car ownership were still low – car densities per 1000 people ranged from 43 in the Soviet Union itself to 200 in Czechoslovakia and 233 in East Germany – and environmentalists saw the successor states as 'uniquely positioned, with extensive public transport systems, to avoid the excesses of auto dependence'. But governments and the European Bank for Reconstruction and Development invested heavily in roads, and in the 1990s car ownership levels doubled.[26]

There were great differences in types of fossil fuel use between higher income developing countries and the poorest countries, and within countries, as shown in China, India, and parts of sub-Saharan Africa.

In China, the context for big changes in energy consumption patterns was the economic policy shift of 1978, towards using market mechanisms more widely and reducing central state control. The economy's overall energy intensity fell by half between 1980 and 1995. A group of US-based researchers identified the main causes of improved efficiency as: state regulations that required the shutdown of small, inefficient electricity plants; changes in ownership and rising fuel prices, stimulating energy savings; and structural change, including a shift to light manufacturing. Industry grew rapidly; therefore greater efficiency did not mean lower consumption. The cement industry, for example, in the early 2000s used only two-thirds of the energy-intensive clinker it had used in 1980 for each tonne of output – but output was five times greater. In the 1990s, efficiency progress stagnated. Many firms, once in private hands, made one-off efficiency gains and avoided investing in efficiency during the 1990s, when energy prices were relatively low. Between 1983 and 2003, conservation investments as a proportion of total energy investments fell from 13 per cent to 4 per cent.[27] Indian industry made some energy efficiency gains in the 1980s, but these were overshadowed by the sharp rise in agriculture of energy use, energy intensities and electricity supply; the latter leapt up by 11 per cent/year between 1970 and 1999.[28]

For urban households outside the rich world, electrification was a significant trend – the most significant change in types of fuel used, a survey of 18 developing countries between 1978 and 1984 found. In Pune, in Maharashtra, India, higher-income households invested in electricity connections and sub-let electricity to poorer neighbours. Between 1982 and 1989, the proportion of the city's households with a fan rose from 39 per cent to 65 per cent; with a TV, from 25 per cent to 68 per cent; and with a refrigerator, from 13 per cent to 23 per cent.[29] Research of middle-class homes in Kerala, India, in the mid 1990s also showed big rises in electricity and fuel use. In one district of Trivandrum, 82 per cent of households had refrigerators, 90 per cent electric irons, 56 per cent washing machines and 39 per cent cars. The number of private cars in Kerala rose by 50 per cent in 1990–95 and doubled again in 1995–2000.[30] For most Indian families, though, these goods remained unattainable luxuries. In 1983, the top 10 per cent of the population accounted for 30 per cent of household energy use, and the bottom 60 per cent of the population for 33 per cent; in 2000, those figures had hardly changed, while other measures of inequality – such as the gap between the rates at which different regions reduced the proportion of people without access to LPG and kerosene, let alone electricity – had risen. The background was an overall reduction of poverty, but widening inequalities between India's richest and poorest, between the cities and countryside, and between states.[31]

A minority of the highest-income households – largely from the 1980s in China, and the 1990s in India – adopted energy-intensive patterns of middle-class consumption modelled on those of rich countries. By the early 1990s, after a decade of market reforms, more than a million Chinese households were in the yuan-millionaire bracket (i.e. earned more than $125,000/year); 4.3 million earned more than 30,000 yuan ($3750); and 100 million earned more than $1,000/year and aspired to new types of consumption. Whereas in the 1970s, China's three great consumer desirables had been a transistor radio, a bicycle and a sewing machine, by the 1980s, the wish-list was similar to Japan's: TV sets, refrigerators and washing machines. By the 1990s, the three desirables were a video, CD player and air conditioning. Phones, privately owned apartments and cars were seen as 'super-products'. These changes of culture were driven by the Chinese government and foreign investors as much as by households themselves. Advertising, which was essentially non-existent in the 1970s, had by 1994 become an industry with $1.2 billion annual turnover.[32] In India, the 'open economy' policy of 1991, designed to encourage foreign investment and market mechanisms, resulted in an explosion of TV ownership and of advertising, for which TV was the main vehicle.

As city dwellers acquired appliances, rural China remained, in many ways, another world. Between 1980 and 2004, refrigerator ownership rates rose from 7 per cent to 90 per cent in towns, and 0.1 per cent to 12 per cent in the countryside; washing machine ownership rates rose from 48 per cent to 96 per cent towns, and 2 per cent to 37 per cent in the countryside. The most significant shift in rural household energy use in the 1980s was a government programme to promote improved biomass cook stoves, with the aim of cutting demand for both biomass and coal, and helping to tackle pollution and health problems associated with older stoves and household coal use. By 1992, 129 million new stoves had been installed.[33]

Consumption patterns in sub-Saharan Africa were very different from those in China and India. Many already-poor oil-importing countries were harder hit by the second oil shock, and took longer to recover, than others. They fell further into poverty. In the Sahel countries,[34] commercial energy consumption, which had in 1973–80 grown on average by 3.5 per cent/year, in 1980–86 fell by 0.5 per cent/year. Poor families turned back from LPG and kerosene to biofuels, which had 'serious implications for further deforestation, desertification and environmental degradation'. By 1990, only 16 per cent of the population of sub-Saharan Africa had electricity access (and many of those were in one country, South Africa), compared to a rate of 46 per cent across all developing countries. In industry, where other countries had in the early 1980s invested in efficiency, the poorest nations struggled to find 'the enormous investment required to effect the replacement of older capital stock', an OPEC researcher found. Even relatively well off Nigeria, Ghana and Liberia experienced a fall in aggregate energy consumption in the decade to 1988.[35]

Energy policies I: enter neoliberalism

In 1981 Jimmy Carter was succeeded as US president by Ronald Reagan, who stood for tax cuts, budget cuts and dismantling state regulation. This meant significant changes in energy policy. Reagan ended remaining price controls on petrol (largely a piece of showmanship, since they were due to expire and prices were falling), loosened control on leasing federal lands and offshore areas for oil drilling, and undermined the EPA and environmental regulation. He slashed research and subsidy support for energy conservation, and for renewable energy sources, while backing nuclear development. In 1981 the carmakers lobbied Reagan to protect them from Japanese competition with import quotas, and in 1985 to undermine fuel standards rules (see p.128): he willingly acceded on both counts. Reagan's assault on environmental regulation, in the name of free enterprise, was the seedbed from which first 'environmental scepticism', and then climate science denial, gained momentum on the right wing of US politics.[36]

Reagan and UK prime minister Margaret Thatcher, who had taken office in 1979, were seen as standard-bearers for neoliberalism, an ideology originating in economic ideas, opposed to Keynesianism, developed in the 1940s and 1950s by Friedrich von Hayek, Milton Friedman and the 'Chicago school'. In the 1970s it had influenced the dictatorships in Chile (from 1973) and Argentina (from 1975) that rolled back social democratic and statist policies. Under Reagan and Thatcher, neoliberalism came to mean a range of policies that supposedly eroded the state in favour of markets, such as privatisation, deregulation and market liberalisation. Neoliberalism is sometimes portrayed as a dogma that made market mechanisms almighty, but actually the withdrawal of the state from the economy only happened in rhetoric. Rather than doing away with state intervention, the neoliberals changed its aims. David Harvey argued that neoliberalism could be seen as 'a utopian project to realise a theoretical design for the reorganisation of international capitalism' or as 'a political project to re-establish the conditions for capital accumulation and to restore the power of economic elites'; this latter function dominated. This political project was given force by the post-boom crisis of profitability discussed above.[37]

Electricity sector reform provides an example of the impact, and of the limitations, of neoliberal policies. In the rich world, the economic crises of the 1970s had left the electricity sector with surplus generating capacity. Large central power stations, constructed on the assumption that electricity demand would keep growing as it had during the boom, turned out not to be needed. In the USA in 1982, the surplus was nearly 40 per cent of total capacity. Nuclear projects, the most expensive, were especially vulnerable to collapse. The Washington Public Power Supply System, the largest electricity supplier in the northwest USA, cancelled four of five nuclear reactors that it had started, and, when the two largest were trashed in 1982, carried out the world's biggest corporate bond default up to that time. (The company was renamed Energy Northwest in 1998.) While nuclear technology struggled, gas-fired power – that allowed much smaller stations and some decentralisation – flourished. US market liberalisation, initiated by PURPA in 1978, forced large generating companies to make room for some cogeneration capacity owned by industrial firms, and stimulated the growth of gas-fired power. (See p. 103.) The limits of neoliberal policy were evident, though, in the billions of dollars worth of state support given to the nuclear industry, in seeking fixes for safety problems, below-market insurance cover, and other benefits.[38]

The US electricity sector had always been largely privately owned, whereas in Chile and the UK, systems had been state-owned, and neoliberal governments now championed privatisation. In Chile, the electricity sector was sold off in the late 1970s, as part of a gigantic privatisation programme, aimed at weakening organised labour and attracting capital. The UK sell-off came in 1989. The principle used there, of 'unbundling' (separating) generation, transmission and distribution – with transmission treated as a natural monopoly and the other spheres as markets – became the standard model used in 1990s privatisations internationally.[39]

Energy policies II: subsidies

The limits to neoliberalism in energy policies were illustrated most clearly by the level of subsidies by governments to energy producers and consumers. Economists had begun researching these methodically in the 1970s; the issue came closer to the centre of public debate in the 1980s. Politicians and civil society groups began to use subsidies provided – rather than speeches made, or claims about the state's withdrawal from markets – as a standard by which to judge energy policies.

Definitions of subsidy, and ways of measuring them, themselves reflect political standpoints and have changed over time. By way of introduction, the reader's attention is drawn to three points. First, the most convincing definitions of subsidies are very broad. Doug Koplow wrote:

> Government interventions encompass a wide range of regulatory, fiscal, tax indemnifications and legal actions. By modifying the rights and responsibilities of various parties involved with the energy sector, these actions decrease (subsidise) or increase (tax) either energy prices or production costs.[40]

Subsidies are 'often viewed primarily as cash payments from a government agency to private businesses or individuals', Koplow continued, such as heating grants to low-income households or research grants to oil companies – but a more accurate definition would include 'any government-provided goods or services, including risk bearing, that would otherwise have to be purchased on the market'. For an oil or coal producer, such services could include anything from favourable licencing terms or tax breaks to remediation of exhausted fields. For an energy consumer it would include any measures that pushed the price below that on the world market.

Second, the distinction between *producer subsidies*, that lower energy producers' costs, and *consumer subsidies*, that lower the prices paid by consumers, should be borne in mind.

Third, where subsidies are estimated by comparing actual prices with reference prices, there are dilemmas about how to set the latter. It is much easier to do this with oil and oil products, which are traded on world markets, than with, for instance, electricity. A methodological question is whether to include in the reference price the cost of those *externalities* of energy production and consumption, such as air pollution and, on a larger scale, global warming. The answer is political: including such externalities highlights governments' failure to ensure those costs are covered. To the cost of government action (fiscal subsidies), reference prices including externalities add the cost of government inaction (failure to compel energy producers and consumers to pay for the consequences of their activity).

The focus of subsidies research in the 1970s and 1980s reflected energy policy discussions at the time. An early report, published by the Ford Foundation in response to the 1973 oil price shock, advised the US government on tax policy. It highlighted fiscal subsidies to US oil companies, in particular the favourable tax treatment of upstream and foreign operations; favourable tax treatment for private electricity companies and tax exemptions for municipally-owned ones; and direct government funding of energy research. (See pp. 80–1.) In the late 1970s, the US Department of Energy made two studies of its own. The first covered producer subsidies: it provided an eye-watering estimate that $217 billion (1977 dollars) worth of support had been given to energy companies between 1918 and 1976. It also showed – at a time when policy discussions were focused on supply alternatives to imported oil – that the bulk of production subsidies, $21.3 billion out of $32.2 billion in 1977, went to oil. A second report on consumer subsidies also concluded that oil, rather than other fuels, was the main beneficiary.[41]

In the 1980s, environmentalist NGOs commissioned research on subsidies as part of their rearguard (and ultimately unsuccessful) action to defend limited US state support for solar power. One report showed that, of the total $44 billion US energy subsidies in 1984, more than $41 billion went to 'mature energy technologies that long ago reached commercialisation', including $15 billion to nuclear and $8.5 billion for oil – compared to $1.7 billion for all renewables. The report showed that in 1979–85 renewables had supplied 100 times more incremental electricity than the nuclear industry, which had received more than $200 billion in federal subsidies. It warned, correctly, that 'the impending collapse of government support' for renewables would undo almost all the progress made.[42]

Research on consumer subsidies focused on a different set of political issues. It confirmed that petrol taxes in the USA and Canada were far lower than in other rich countries, reinforcing car dependency. But it focused more attention on developing countries, where governments had held the prices of petrol, kerosene and other oil products way below import prices during the second oil shock. Oil importing countries' governments had allowed prices to rise further than had oil exporter governments, research commissioned by the World Resources Institute showed. It also highlighted the extent of coal subsidies in China and India, which together accounted for 70 per cent of developing-world coal consumption. In China, coal prices were one quarter of the average world price and less than 40 per cent of long-run marginal costs, spurring 'excessive demand and frequent supply shortages', which led in turn to industrial capacity being stood down due to fuel shortages.⁴³ In the context of China's market reforms, coal prices were increased gradually throughout the late 1980s and 1990s.

Road transport subsidies, usually in the form of government support for building roads and parking spaces in preference to other transport infrastructure, are a gigantic stimulant of oil products consumption – but are usually counted separately from fossil fuel subsidies. The body of research on them is limited. Norman Myers, in a general survey of subsidies published in 1998, argued that road transport subsidies were the world's largest type of 'perverse subsidy', and larger than subsidies to fossil fuels, agriculture, water or fisheries.⁴⁴

Subsidies, like government policies in general, reflect relationships of power and wealth. Producer subsidies, particularly in rich countries, often reflect state support for powerful industrial sectors. Consumer subsidies, especially in oil-importing developing countries, often reflect social compromises between governments and both industrial capital and households. In the 1990s, the international financial institutions – which had rarely interfered noticeably in the relationship between rich country governments and energy producers or consumers – often made the reduction of consumer subsidies a condition of loans, and of other forms of support, to developing country governments. This in turn fed into sharp conflicts between energy consumers, especially the urban poor, and governments who they perceived to be victimising them with fuel price increases.

The discovery of global warming

The discovery of global warming was a triumph of late twentieth-century science, no less momentous than mapping the human genome or advances in understanding the universe's origins. The collective efforts of thousands of meteorologists, paleoclimatologists, oceanographers and others produced, by the mid-1980s, a consensus that warming was underway; that the greenhouse effect was probably (that soon became 'certainly') the major cause; and that that in turn was due to human activity, first and foremost fossil fuel combustion. Christophe Bonneuil and Jean-Baptiste Fressoz have argued against 'grand narratives of awakening, revelation or arousal of consciousness' in the 1980s about the damage done by society-nature interactions; society's potential to do serious damage to the natural world, and to itself, had long been there for anyone who wanted to see it – and it was seen, at least since the eighteenth century.⁴⁵ Nevertheless, the discovery of global warming made precise the character of a particular danger that had hitherto been unclear. By the end of the 1980s, political leaders accepted – in some cases, as though having been dragged, kicking and screaming – the need for a coordinated international response.

Most climatologists had by the late 1970s accepted that human activity could change the climate, but there was no consensus on how (e.g. whether cooling, warming, or atmospheric pollution effects would be dominant). The first international conference on the issue, organised in 1979 by the World Meteorological Organisation (WMO) and the UN Environment Programme (UNEP), proposed only more systematic exchange of information. In the early 1980s, researchers using computer models that reproduced climate and weather phenomena (general circulation models) made progress. The models had from the 1960s produced educated guesswork, but by the 1980s, thanks to advances in computing, were providing accurate short-term weather forecasts. Two research groups that crunched data on long-term temperature trends now concurred that the atmosphere had warmed since the mid-1960s by 0.2°C. A consensus formed that mid-century cooling in the northern hemisphere had been an aberration, that the most likely cause of warming was the greenhouse effect, and that it would be possible to ascertain this from physical observation by the end of the century.[46]

A second important breakthrough came from paleoclimatologists using deep drilling technology to recover prehistoric ice cores. They devised ways of measuring the chemical composition of bubbles trapped in the cores to build a long-term chemical picture of the atmosphere. Readings from the Soviet Vostok station in Antarctica, which in the 1980s recovered samples going back over 400,000 years, 'tipped the balance in the greenhouse-effect controversy, nailing down an emerging scientific consensus' on the causal role of carbon dioxide, Geoffrey Weart, the historian of climatology, wrote. The ice cores contained 'disturbing' new evidence that the climate could change far more rapidly than scientists had previously thought (i.e. over tens or hundreds, rather than thousands, of years). By the late 1980s much research was focused on disentangling the relative importance of various greenhouse gases, including CO_2, methane, and aerosols and sulphates, and other possible causes of climatic change. The role of oceans, and their ability to retain heat in some periods and release it in others, became better understood. Oceanographic data was incorporated into the computer models.[47]

State and political bodies took notice. In the USA, the National Academy of Sciences was asked by government to investigate. In 1983 it expressed 'deep concern' about possible anthropogenic warming, and recommended further research. In the same year a report by the US EPA was less sanguine, 'raising the spectre of a world on a collision course, between the need for energy derived from coal and a global warming of potentially catastrophic proportions', as two EPA officials recounted.[48] International dialogue between scientists, diplomats and political leaders was formalised by the WMO and International Council of Scientific Unions, who, in October 1985, together with the UNEP, organised a conference at Villach, Austria. Scientists in attendance from 29 countries supported the conclusion that a 'significant' global warming, caused by the greenhouse effect, was likely during the first half of the twenty-first century, and scientific-political cooperation was needed.[49]

Such warnings entered an international political discourse that had already accepted the impact of human economic activity on the natural environment as its concern. The UN had in 1983 appointed a World Commission on Environment and Development (known as the Bruntland commission after its chairman). Its report, issued in 1987, referred to the Villach meeting and asked of global warming: 'How much certainty should governments require before agreeing to take action?' If they waited until physical evidence of global warming arrived, it might be too late for countermeasures to be effective. 'Internationally agreed policies' to cut greenhouse gas emissions should be developed, along with monitoring, research, and adaptation strategies for rising sea levels.[50]

In 1987, too, came the discovery of a gaping hole in the layer of ozone around the earth, and the rapid conclusion of an international agreement to reduce the production of chlorofluorocarbons that had caused it (the Montreal convention). This indicated both that human economic activity could cause drastic climate change, and that international cooperative action could potentially produce remedies. In 1988, the Intergovernmental Panel on Climate Change (IPCC), the scientific-political body that would play a key role in UN climate negotiations, was formed. A World Climate Conference convened in Toronto, Canada, in the summer of that year. The scientists and diplomats present proposed that CO_2 emissions be reduced by 20 per cent of their 1988 level by 2005, by means of strict targets aligned with energy policies – and were optimistic that political leaders would act.[51]

These hopes would be dashed. At the UN Conference on Environment and Development (the Rio summit) in June 1992, the UN Framework Convention on Climate Change (UNFCCC) was signed, but without any of the targets or coordinated energy policies for which the scientists had hoped. The retreat of Keynesian state direction of the economy in the 1980s; the deep prejudice against regulation, underpinned by neoliberal ideologies, that embraced governments and state bureaucracies; and the hostility of coal and oil producers to regulation; all played their part. In 2005, world CO_2 emissions would be not 20 per cent lower than the 1988 level, as had been proposed in Toronto, but 35.3 per cent higher.[52]

9
The 1990s: shunning the global warming challenge

International political action to deal with global warming began with the convention adopted at the Rio Earth Summit in 1992 (the UNFCCC), and continued with the Kyoto Protocol of 1997. This process failed to reduce fossil fuel consumption, the main source of greenhouse gas emissions. By 2000, fossil fuel use was 16.1 per cent higher in the OECD, and 10.3 per cent higher outside the OECD, than it had been in 1990.[1] Energy policies had been shaped during the 'roaring nineties' not by the Rio decisions, but by the imperatives of economic growth. Rich-country governments not only underpinned fossil fuel consumption growth with tax and subsidies policies, but also worked with international financial institutions to liberalise energy markets, particularly electricity, in ways that cut across or flatly obstructed conservation. The challenge in writing the history of these events is to understand exactly how political, social and economic forces combined to produce a disaster of this magnitude.

Global warming: the political reaction

The IPCC's first assessment report, published in 1990, acknowledged 'significant uncertainties' in the research of global warming's effects (e.g. sea-level rise, more volatile weather, and other dangers to people and agriculture), but underlined that enough was known 'to begin adopting response strategies', including 'efficiency improvements and conservation in energy supply'. There was less doubt about the role of fossil fuels as the single largest cause of human-made warming: the report attributed 70–90 per cent of CO_2 emissions, and therefore 38–54 per cent of the greenhouse effect as a whole, to energy production and use.[2]

The most powerful countries agreed that a treaty was needed. The US administration, headed by president George H. W. Bush, dominated its preparation, particularly since the second post-war superpower, the Soviet Union, was in terminal decline. The US strategy was managed by John Sununu, White House chief of staff, a climate science denier who believed environmentalists sought to impose de-growth via the treaty. US priorities were to ensure that there would be no binding targets or timetables for cutting greenhouse gas emissions, and that any policies agreed would support economic growth. The idea was to 'use market approaches in setting environmental policies', US officials commented in retrospect. Regulation was out. European governments, in contrast to the USA, favoured deadlines: the European Commission advocated stabilising CO_2 emissions by 2000.[3]

The sharpest conflict, though, was between the USA and large developing countries, whose representatives argued that – since rich countries were 'mainly responsible for causing climate change', in the words of a Chinese document – they were also mainly responsible for dealing with its consequences. US diplomats tried to twist developing country representatives' arms by diverting attention away from CO_2 on to deforestation (which aggravates the greenhouse effect). In a parallel dispute, non-governmental organisations (NGOs) based in the Global South clashed with some rich-country environmentalists whose research fed the US narrative. A report by the World Resources Institute exaggerated deforestation as a cause of global warming, thereby detracting from rich countries' responsibilities, Indian activists argued.[4]

The treaty, which was signed at Rio in 1992 and entered into force in 1994, conformed to US demands. It declared the aim of stabilising greenhouse gas concentrations in the atmosphere 'at a level that would prevent dangerous anthropogenic interference with the climate system', without saying what that level was. Rich countries made non-binding commitments to return to 1990 levels of emissions by 2000. The energy researcher Jose Goldemberg, then Brazil's environment secretary, said the 'true measure' of rich countries' commitments were the laughably small contributions to the Global Environment Fund set up at Rio – $50 million from the USA, for example.[5]

The IPCC's second assessment report, published in 1995, outlined more accurately the likely effects of global warming, forecasting for the twenty-first century a 2–3.5° rise in average temperatures, sea-level rises of 50–95 cm and other frightening 'surprises'. This gave impetus to renewed arguments by countries including the UK, Germany and Denmark – and by the Democratic US administration headed by Bill Clinton, who replaced Bush as president in 1993 – for greater collective commitment to emissions reductions. There was a blistering counterattack from the climate science denial movement, which had since the late 1980s been growing on the right wing of US politics. The Global Climate Coalition, a lobbying group set up in 1989 to undermine the Rio process, with funding from Exxon, Shell, BP, Ford and Chrysler, worked with the US National Coal Association, Auto Manufacturers Association and others, to forestall agreement on binding targets. Exxon, despite employing climate scientists who shared the consensus understanding of fossil fuels' causal role in global warming, used big-circulation publicity material to bolster the denialists' false claims by fostering doubts about the science.[6]

As international negotiations reached their climax in 1997, in the run-up to the Kyoto conference on climate, this coalition broke down. BP's US subsidiary withdrew from it in 1996, to be followed in 1997-8 by Shell, Sunoco and Texaco. But much damage had already been done. In July 1997 the US Senate voted by 95-0 against binding commitments. The Protocol agreed at Kyoto in December 1997 put legally binding limits on rich countries' emissions of CO_2 and five other greenhouse gases, amounting to a 5 per cent reduction on 1990 levels, to be achieved by 2008-12. But the agreement was never ratified by the USA, the largest emitter to which it would apply; and Canada, the third largest, ratified it in 2002 but withdrew in 2011. Market mechanisms, enabling the trade of permits to emit greenhouse gases, were set up, to allow countries to add supposed improvements elsewhere to their own greenhouse gas accounts. Russia, Ukraine and other former Soviet countries were excused from taking action by another accounting trick: since 1990, the guideline year against which emissions were measured, they had suffered the biggest ever peace-time economic slump, which had reduced fuel consumption by between a quarter and a half. China, now the second largest emitter, led the efforts by developing countries against any current or future commitments.[7]

Economic growth and energy policies

When looking back at the Rio treaty's lofty declarations, it should be borne in mind that global warming was not the first concern of the politicians and civil servants who produced it. It was still less important to the heads of international financial institutions, energy company executives, economists and others, who took myriad other actions that shaped fossil fuel consumption trends. To the extent that they saw what they were doing as progress, that usually meant economic growth. Ideas about sustainable development, to which the Rio treaty made approving reference, were also folded into assumptions about growth.[8]

The treaty was adopted in the midst of the ragged, uneven boom of the 'roaring nineties'. The former Soviet countries were mired in slump, and Japan was jolted by a market collapse that destroyed half the Nikkei share index's value in 1989-99 – but the USA and Europe boomed: in that decade, their stock exchanges rose by four times and 2.5 times respectively. Continuing globalisation was part of the story. Capital flowed to east Asia, in enormous quantities. Private financial flows to developing countries rose from $44 billion in 1990 to $244 billion in 1996. New technology (including the establishment of the world-wide web in 1993) made a difference: in the USA, the dot-com boom became a market bubble.[9]

During this boom, neoliberalism held sway politically. The Soviet collapse proved, in the ideologues' minds, that their offensive against state provision was justified. Globalisation would be enhanced by deregulation and liberalisation of markets, and privatisation. The international drive to lower barriers to capital and goods led to the formation of the World Trade Organisation (WTO) in 1995. In the IMF and the World Bank, which influenced developing country governments, the 'Washington consensus' – a menu of free-market strategies, from trade liberalisation to privatisation – was all-powerful. In the EU, the new objective of monetary union, to support a unified, open market, drove deflationary policies.

Labour and social movements resisted neoliberalism in myriad ways. In rich countries, beleaguered trade unions fought to defend the 'social wage' (health, education and welfare services). Outside the rich world, newly urbanised people battled hardships born of neoliberal dogmas. On the other side of the coin, anger at governments' failure to act on global warming came not only from rich-world protesters, but also from the 'environmentalism of the poor', manifested in bitter clashes with energy and mining companies who saw liberalisation as a green light to exploit mineral resources with scant regard for those who lived in the vicinity. Resistance to oil companies' activities in Nigeria became a prominent example.[10] But these movements were disparate: they obstructed and confounded neoliberal governments' actions, but were not strong enough to reverse the tide.

Against this background, energy policies were aimed not at conserving, or shifting away from, fossil fuels, but at liberalising and expanding markets. There was a wave of electricity deregulation and privatisation. (See pp. 143–7.) By the turn of the century, liberalisation was 'by no means complete', though, Michael Pollitt concluded from a comparative study. Oil and gas production companies were privatised in a range of countries, including Russia – but not in the Middle East. Gas distribution companies that supplied end users were hard to privatise: they were more than 50 per cent state owned in 16 of 39 leading countries Pollitt surveyed. In coal, Pollitt wrote:[11]

> [L]iberalisation has taken the form of the running-down of subsidised domestic coal production in many high cost countries (e.g. Spain, UK, Germany and Poland) with some privatisation of residual coal assets (e.g. in 1994 in the UK). The result has been a significant increase in world trade in coal from countries with mostly unsubsidised, privately-owned coal assets (such as Australia and South Africa). [...] Even the large Chinese coal industry has been subjected to significant liberalisation, with many enterprises being privatised and/or decentralised down to local municipalities.

While liberalisation sought to open up international energy markets, many governments continued to provide huge subsidies to domestic fossil fuel producers. And despite energy prices being relatively low in the 1990s, governments, both in the rich world and outside it, also subsidised consumers, by setting prices of oil products or gas below world market levels and/or by taxing fossil fuels leniently. Research on the subsidies and the way they contradicted climate policies began to be supported by international institutions. High energy prices had 'the potential to reduce energy consumption and encourage more efficient energy use, without sacrificing economic growth', an IEA study argued in 1993 – but 'heavy fuel users, heavy polluters and fuels with a high pollution potential are seldomly and/or lightly taxed'. In some rich countries, the greater the damage from a fuel's emissions, the less tax was paid on it. Australia and Germany had no tax on burning coal, and in the USA, 'virtually no taxes are levied on energy use by stationary sources [i.e. in industry and electricity generation]'.[12]

Economists at the World Bank estimated in 1992 that global energy subsidies were running at $230 billion/year, and argued that removing these would reduce global carbon emissions by 9 per cent. Their research used the price gap approach, that measures the difference between world prices and end-user fuel prices – essentially, the gap between a nominal world price and what a Russian householder, an Indonesian taxi-driver, or a Canadian petrochemicals plant, would pay – and thereby *excluded* many significant subsidies to fossil fuel producers. (See pp. 133–5.) Using the same method, the IEA estimated fossil fuel subsidies were in 1999 worth, on average, 11 per cent of world prices in China, 14 per cent in India, 33 per cent in Russia, 80 per cent in Iran, and so on.[13]

The IMF and World Bank encouraged – and in the aftermath of the Asian financial crisis, compelled – developing-country governments to reduce or remove these subsidies. Plenty of research showed that such subsidies (e.g. cheap diesel for an Indonesian taxi driver or cheap heat for a Russian household) rarely and inadequately benefited the final consumers; hardly ever benefited the poorest people in developing countries; and that the money spent on them might have gone far further in benefiting the poorest had it gone e.g. to health care.[14] Nevertheless, attempts to remove subsidies were often fiercely opposed by social movements who saw them as tangible threats to benefits today, in the name of an intangible better tomorrow. (See pp. 148–9.)

These efforts to reduce developing-country consumer subsidies fitted well in neoliberal policy frameworks. Reforms to rich-country taxes that might adversely affect their fossil fuel producers, or profligate consumers, did not – as was demonstrated by the short-lived attempt to introduce a wide-ranging energy tax in the USA in 1993. Lawrence Summers, then World Bank chief economist, made the case for a carbon tax; president Clinton proposed a so-called 'btu tax' that was a close approximation. (Btu stands for British Thermal Units of energy.) The coal industry and climate science deniers first watered it down by pressure on the administration, and then secured its defeat in the US Senate.[15] Such proposals, hardly radical or anti-capitalist but supported by US Democrats and European social democrats alike, could gain no traction in the heyday of neoliberalism.

Governments and corporations turned their backs not only on proposals to encourage energy efficiency with tax and subsidy reform, but even on commitments made at Rio, to monitor emissions and use the best available to technology to cut them. Sixteen years later, in 2008, the IEA complained that 'data are [still!] particularly poor for the iron and steel, chemical and petrochemical and pulp and paper sectors' – that is, the most fossil-fuel-intensive industries; that data breakdowns of fuel use by different industrial processes did not exist; and that 'company-based data are subject to confidentiality problems associated with anti-trust legislation'. In a report published 23 years after Rio, tracking transport, buildings and electricity generation as well as industry, the Agency complained that 'limited data availability and poor data consistency' continued to obstruct conservation strategies.[16]

Electricity reform

The most significant impact of energy sector liberalisation was on electricity systems. In much of the rich world, these had been partly or wholly state- or municipally-owned, and regarded by much of society more as essential infrastructure, or a means of social provision, than as a business. Neoliberalism questioned these assumptions. The UK electricity reform in the 1980s was followed in the 1990s by various combinations of privatisation, restructuring and the introduction of competition in Italy, Germany, France, Sweden, Finland, Australia and New Zealand. The European Commission's 1996 electricity directive aimed at creating competition across national borders. In the USA, where electricity generation was largely privately owned but monopolised, the federal regulator in 1996 required transmission operators to open up their networks to electricity generated by competitors.[17]

There was a technological aspect to reform. The arrival of combined-cycle gas turbines made possible the construction of small (e.g. 300 MW) gas-fired power stations, at one third or less of the cost of a typical coal station. (See pp. 29–31.) The idea of a competitive market for generation, based on such stations, was seductive. Walt Patterson wrote:

> Until the 1980s, the ruling assumption was always that a better power station meant a bigger power station. The assumption was questionable even then. At the turn of the millennium it had been comprehensively discredited. [...] The gas turbine option, with its low capital cost and rapid construction time, has made life even more difficult for nuclear power, with its high capital cost and protracted construction time.[18]

The really controversial aspect of electricity liberalisation was its application – or, more accurately, imposition – in developing countries by their governments, international financial institutions and rapidly growing multinational companies. The latter included AES of the USA, which by 2002 had operations in 32 countries; more than two dozen US and Canadian utilities; and European energy firms including National Power, PowerGen, E.ON, EdF, RWE and Vattenfall. Firms with origins in other businesses, such as ABB of Sweden, ExxonMobil, Bechtel and Foster Wheeler, jumped in. The largest, Enron, invested overseas billions that it had made through trading electricity in the USA before its scandalous collapse in 2001. European and US deregulation, purportedly aimed at encourage competition, concentrated assets in such companies' hands, helping to fund their international expansion.[19]

The companies' most important ally was the World Bank, which, inspired by the 'Washington consensus', pushed a 'standard model' of electricity market reforms across the developing world. The Bank would 'aggressively pursue' commercialisation and corporatisation of electricity sectors, it declared in a 1993 policy paper; lender countries would be required to encourage private investment, and establish transparent regulatory processes. Electricity generation assets were most often, and most quickly, privatised. Typically, the multinationals entered developing countries and set up IPPs that sold electricity to (often state-owned) utilities. The latter focused on the relatively costly and difficult business of distributing the electricity to customers and collecting payment. In theory, the magic of the market would address developing-country networks' very real problems, including chronic under-investment, growing demand, and gross inefficiencies, both technological and bureaucratic. Pricing regulation would both tackle inequitable subsidies and, ultimately, result in lower tariffs. In practice, the multinationals, supported by the Bank and its consultants, often negotiated long-term contracts for their IPPs' wholesale electricity that left the significant risks with the state, while steering clear of the long, arduous business of improving underfunded distribution systems.[20]

India is the largest developing country where the World Bank's formulas were applied, with enthusiastic support from the government, which in 1991 wholeheartedly adopted neoliberal economic policies. At that point, about two-thirds of India's electricity was generated by the SEBs, owned by state governments. Most of the remainder came from three (federal) state-owned corporations that owned thermal, hydro and nuclear stations. The SEBs also undertook most transmission and distribution to final consumers. While there were some shortages of generation capacity, the SEBs' most chronic problems were financial. They managed complex systems of subsidies to various categories of consumers. They struggled with electricity losses estimated at up to 40 per cent in some places, from aging infrastructure, unmetered consumption and theft.[21]

The electricity law amendments of 1991 sought to attract international investors into generation, with generous incentives (including a guaranteed 16 per cent return on equity investments) and a fast-track project approval process.

Initial enthusiasm soon drained away: of 190 applications, only 15 projects were approved and even fewer implemented. Public discussion centred on Enron's notorious Dabhol plant in Maharashtra. The firm, strongly supported by US diplomats – and the local police, who brutalised its opponents – negotiated an electricity sales contract at unsustainably high prices. That helped to plunge the state's SEB into $300 million worth of debt before it defaulted. Since Enron folded in 2001, the plant has been idle.[22]

Alongside the stuttering privatisation programme for generation, the World Bank sought a shop window for its model of unbundling, with regulated transmission and distribution. It focused its efforts on Odisha (formerly Orissa) state. The researchers Navroz Dubash and Sudhir Rajan found that the reforms were 'single-mindedly focused on financial issues and privatising the sector'; the World Bank and other international donors were 'obsessed with removing subsidies and increasing tariffs'; and to attract investment, they changed the rules, by transferring most financial liabilities to the publicly owned transmission company in an attempt to privatise distribution. Privatisation revenues were not returned to the system, and there were big tariff increases but no improvement in service. Nevertheless, Odisha was used as a model for reform attempts in other states. Research on the reform concluded that it largely failed, both on its own terms (the investors did not produce substantial new generation capacity) and more generally (it found few answers to SEBs' financial problems, weakened key institutions, and did not improve the quality of service or substantially contribute to electrification).[23]

These problems – with unsustainably favourable deals for IPPs, and with transmission and distribution reforms running into economic and social obstacles – were repeated in many countries. In the Philippines, 45 new power stations were built by foreign investors, but power purchase agreements that left all the risks with the state-owned National Power Corporation brought it to its knees after 1997. There were allegations of corrupt patronage in contract awards. The contracts were renegotiated; electricity shortages remained. Events in Thailand took a similar path, with the Prachuab Kirkhan IPP, in particular, being stalled due to public protests against anticipated environmental impacts. In Kenya, UK-, Spanish- and Malaysian-owned IPPs concluded deals to sell electricity at rates far higher than those charged by the parastatal generator, Kengen; contracts were awarded in breach of tendering procedures; frictions were compounded by misallocation of rural electrification funds; and in 2004 the government decided to phase out IPP contracts.[24]

In the 2000s, the markets' own logic – most dramatically through the California electricity crisis, and the collapse of Enron, in 2001 – would confound the reform process. (See pp. 160–1.) But even before that, reform barely ever attained its advocates' stated aims. Reforms were partial, fragmented, and moulded as much by national circumstances as by economic policy prescriptions. Developing-country governments facing shortages of generation capacity started by encouraging IPPs, but often went no further; 'with few exceptions, market forces operate[d] only at the margins' after reform attempts, David Victor and Thomas Heller concluded from a comparative survey. These 'hybrid systems' were not aberrations, but the most likely outcome. Research of reforms in sub-Saharan Africa, commissioned by the World Bank, showed that by 2006, of 24 countries surveyed, 'nowhere' resembled the Bank's 'standard model'. Most countries had hybrid power markets where a state-owned utility remained dominant. There had been $2 billion worth of investment by IPPs, who generally contributed reliable supply to national grids. But these were exceptions, which had added a mere 3 GW of capacity, a drop in the ocean of sub-Saharan Africa's needs. Where energy companies had leased systems, or been awarded contracts to manage them, there had been 'a high rate of disappointment'. Independent regulation had made little progress.[25]

Research by economists and political scientists, including some who embraced much of the logic that underpinned the reforms, found, broadly, that global capitalism had been served, rather than the people who use electricity; that corporations, rather than their employees, had benefited; and that the reforms had cut across global warming policies. Navroz Dubash argued that the reforms were 'driven more by ideological considerations than by evidence of the benefits of restructuring'. The experience had shown that 'left to their own devices, markets are ill-equipped to address equity considerations in access to electricity or prices', and 'fail to internalise environmental impacts'. The reforms had been bad for the societies in which they were conducted, economists who surveyed 17 non-OECD Asian countries concluded. They found 'a tension between economic and welfare impacts'. Reform measures associated with 'positive economic growth' had 'a negative effect on welfare indicators' such as the Gini coefficient that measures income inequality.[26]

As for the threat of global warming, Dubash and his colleagues found 'little political commitment to promoting sustainable development through electricity sector reforms'. In five of six countries studied (Argentina, India, Indonesia, Bulgaria and Ghana), 'closed political processes and politically powerful groups constrain[ed] attention to sustainable development'. IPPs 'locked [India and Indonesia] into large generation plants' and this 'undermined efforts at energy efficiency and committed utilities to buying electricity at uncompetitive prices' – even when cheaper supplies were available. By attracting capital only to generation, 'IPPs have potentially negative environmental implications, since they skew incentives towards new generation and against meeting electricity needs through greater efficiency'. They had 'forced use of high-cost power over lower-cost power already available'. There had been 'little effort' to understand the danger that standard reform procedures such as unbundling could confound demand management (i.e. efficiency and other measures to reduce demand).

The World Bank had treated demand management as a box to be ticked and not as a priority.²⁷ It was a vivid example of how aims associated with economic expansion trumped the Rio treaty's aims.

The Asian crisis

In 1997, Asian stock markets and currencies crashed, and billions of dollars of capital fled. A bubble, driven by capital market liberalisation and rapid economic growth, had burst. The underlying causes, economists reckoned, included over-accumulation of capital in the preceding period and excess capacity in the world economy of intermediate products ranging from steel and chemicals to computer chips. Indonesia was worst hit: economic output per capita fell by one seventh in 1998. Financially, turmoil raged in Thailand, the Philippines and beyond.²⁸ World money markets shuddered. In August 1998, Russia defaulted on some types of state debt, and its currency collapsed to one third of its previous value.

For energy markets, the immediate consequence was that oil, too, was in oversupply. World oil prices, which had been falling slowly since 1990, hit their lowest level since 1973 – around $12/barrel in 1998. Natural gas, and to some extent coal, followed. Such prices were 'unlikely' to have incentivised energy efficiency, the IEA commented. The price collapse triggered a wave of oil company mergers, the industry's biggest restructuring since the break-up of Standard Oil in 1911. The 'seven sisters' of the 1970s, which had been reduced to six in 1985 when Chevron swallowed Gulf, became four (Exxon and Mobil merged; Chevron and Texaco merged; Shell survived; and BP grew by hoovering up Amoco, Arco and Burmah-Castrol). Further mergers produced Total-Fina-Elf and Conoco-Phillips.²⁹

Across the developing countries, capital quit the electricity sector: foreign direct investment in non-OECD electricity reached a peak of $40 billion in 1997, crashed to $12 billion in 1999, and, after a brief uptick, fell even lower in 2002. But that was not the only effect. South Korea, Thailand, Indonesia and the Philippines had taken macro-economic stabilisation loans from the World Bank, on condition that they pushed through electricity sector reform; they now had to try to accelerate it under conditions of economic downturn and hardship. Moreover, where foreign companies had set up IPPs with generous power purchase agreements denominated in dollars, state-owned utilities – who collected revenues in local currencies that had sunk in value – suddenly found themselves struggling to pay. Contracts were broken, or renegotiated, in Thailand, the Philippines and Malaysia.³⁰ Consumers at the end of this chain of payments, whose living standards had been crunched by the global market, felt disinclined to pay swollen bills.

China's expansionary budget and monetary policies had protected its economy from the worst ravages of the 1997 crisis. But industry, and the coal mines, suffered from the same over-capacity as the world economy in general. Restructuring of some energy-intensive industry had already begun, under the ninth Five Year Plan (1996–2000), and 84,000 small enterprises were closed in an effort to reduce urban air pollution. The 1997 crisis turned coal oversupply into something like a glut. The government ordered the closure of loss-making state mines and 25,000 small, dangerous private ones (although local authorities sometimes kept them open). Coal quality improved, in a buyers' market, and the effect of long-term policies on energy efficiency, and on corporatisation and partial privatisation, fed through.[31] The ground was being laid for the biggest coal-fuelled economic boom in history, in the 2000s.

Social protests and refusals to pay

After the 1997 crash, Indonesia, whose finances were hit hardest, negotiated bailout arrangements with the IMF. For the Fund's officials, this was an opportunity to deal with what they saw as non-market fiscal policies. Top of their hit list were energy subsidies to consumers, such as controlled prices for kerosene, diesel and other fuels, sold mostly by state-owned firms to households and businesses. The Fund demanded, as a condition of its support package, that these be phased out. In May 1998, as a direct result, prices rose by more than two-thirds. The population reacted with widespread and violent protests. Muhammad Suharto's authoritarian government, which had ruled Indonesia since 1967, revoked the price increases, but then collapsed. Hard bargaining between the IMF, successor governments and the population continued; in 2005, protests erupted again when the price of diesel doubled, and of kerosene nearly trebled. This conflict was symptomatic of a much wider trend. In Ecuador in 1998, increases in cooking gas, petrol and diesel prices were scrapped in response to street protests. In Nigeria in 2000, a 50 per cent increase in petrol prices triggered civic unrest and was revoked. The problem was a long-standing one, IMF officials acknowledged: civic resistance to fuel price rises had brought down the government of Jordan in 1989, and protests against the removal of maize meal subsidies triggered an attempted coup in Zambia in 1990.[32]

The wider context for such protests was the unprecedented pace of developing-world urbanisation in the last quarter of the twentieth century. A century earlier, new urban populations in Europe had aspired to little more than coal for heating and cooking. But those now arriving in Asian and African cities expected that they would have access to ubiquitous electricity, kerosene and motor vehicles, whether for home consumption or to earn a living. At the same time, capital was embracing neoliberalism as its means of social control. Of course the urban poor could have electricity and fuel, and the freedoms these implied – but they had to be paid for, at prices linked to those on world markets. Against this political economy, that was written in to international institutions' development policies, a moral economy took shape that saw minimum supplies of cheap fuel and electricity as urban residents' human rights. (The idea of moral economy was advanced by E. P. Thompson in his study of the world-view of eighteenth and nineteenth century working people.[33]) International development literature pointed out that subsidised fuel and electricity do not benefit the poorest, especially rural, people. But for hundreds of millions of urban working-class families, this was irrelevant. Social conflicts over fuel and electricity prices were, for them, battles against the encroachment of wealth and power on their lives, not a competition with the rural poor.

Social protests over electricity prices in South Africa were among the most organised. In the 1990s, millions of urban households were electrified, but tariffs were seen by communities as unreasonably high. (See p. 120.) Payment discipline was scorned. Eskom, the state-owned electricity utility, introduced prepayment technology and cut-offs – of which there were an estimated 9.6 million in the eight years to 2002. As household tariffs rose – staying higher (up to double) than those paid by the large industrial consumers who used most of South Africa's electricity – social conflicts multiplied. Black township residents had under apartheid used, as a means of protest, boycotts of payment for the minimal municipal services they received; such collective actions were revived. In 1997 residents in the Tembisa township sabotaged electricity meters; the council responded by deploying private security firms, and even the army, to protect infrastructure. Split meters enabling remote cut-offs became common. Residents found new ways to access electricity, including by engineering additional connections. In 2001 in Alexandra, a Johannesburg township, only 16,000 of the 80,000 households using electricity were registered to do so.[34]

In 2000 the ANC government introduced a Free Basic Electricity programme that supplied a limited wattage to poor households with no arrears. But in Soweto, a large Johannesburg township that had been a centre of anti-apartheid resistance, a combination of arrears and inability to pay resulted in cut-offs running at about 20,000 households per month. A community Operation Switch On campaign began reconnections as a collective action, superseding illegal, individual connections. The government felt compelled to intervene, and Eskom, the local authorities and campaign leaders eventually came to an uneasy agreement to scrap arrears.[35]

In Russia and other former Soviet countries, relations between urban residents, government and electricity suppliers were configured differently. Electricity, gas and district heating had been provided since at least the 1970s, in exchange for a negligible fraction of household earnings. After the Soviet Union collapsed in 1991, and Russia and other successor states plunged into recession, living standards fell steeply. In the resulting social crisis, household electricity, heat and gas were stabilising factors. The IMF-backed fiscal policies of the early 1990s led to an acute shortage of roubles and partial dollarisation of the economy. Non-payment and barter were ubiquitous. Fuel bills went unpaid. After the 1998 financial crisis, and especially from 2000 as oil prices rose, the economy began to recover. In Russia, prices of gas and electricity for industry were raised, but supply to households remained heavily subsidised. The government saw low prices as a way of avoiding social discontent. A wave of protest that greeted reforms to welfare payments in early 2004 redoubled the government's conviction that retail gas and electricity prices had to be controlled. They have been tightly regulated, at a level far below industrial prices, up to the present.[36]

In some urban settlements in developing countries, that had neither the history of basic amenities afforded by the Soviet Union nor South Africa's level of community action, residents endeavoured to access electricity without paying. In the Brazilian favelas – urban slums that grew spontaneously without support from municipal authorities – hundreds of thousands of households have been supplied this way, some for years and even decades. Utilities have used successive generations of metering technology to try to collect payment. ANEEL, the national electricity company, estimated in 2012 that 13 per cent of *all* electricity generated was stolen in informal settlements; in Amazonas state, 30 per cent. In 2006, the Brazilian subsidiary of AES, and the US state aid agency USAID began a programme to commodify the electricity supply, using smart meters; 'affordability' turned out to be the key problem.[37] In Delhi, India, slum dwellers were in legal limbo in the 1990s and could *only* access electricity illegally; when distribution was privatised in 2002, one company found that three-quarters of its customers did so; the city authorities then decided to provide electricity to all areas regardless of their legal status.[38]

In Casablanca, Morocco, shantytown dwellers had traditionally sought to steal electricity from a state-owned supplier that refused to connect them. After privatisation in 1997, Lydec, a private company, took over electricity supply; it insisted on installing legal connections rather than face the financial losses, dysfunction, and dangers of electrocutions, caused by illegal ones. Whereas the state-owned supplier had turned a blind eye to 'electricity piracy', Lydec energetically collected payment, and cut off those who refused it, the researcher Lamia Zaki found. Residents responded with sabotage and violence against company staff. Eventually, the 'pirates' who organised theft on behalf of residents negotiated contracts with the company, under which electricity was sold to blocks of between 20 and 50 dwellings, with a community representative in charge of paying the communal bill.[39] Residents interviewed by Zaki explained their motivation:

Zahra: Are we not human? Are we not citizens too? What about human rights? Now, we've at least learnt a thing or two, and if we aren't given our rights we'll take them. Everyone has the right to light. They've even got light bulbs in the countryside.

Hassan: I am a citizen just like the owners of villas and big cars. I'm a Moroccan too, and I take my rights: I have a right to electricity, and I take it, because I'm a human being, and not an animal.

After the oil price surge of the 2000s and the 2008–9 financial crisis, which redoubled hardship for hundreds of millions of the world's poorest slum dwellers, the drive to commodify electricity gained support from some rich-world academics. The International Growth Centre, founded by Jonathan Leape of the London School of Economics and Michael Greenstone of the University of Chicago, and supported by Oxford development specialist Paul Collier, dedicated itself to 'looking for ways to encourage people to pay for electricity' in Bihar, one of the poorest Indian states, where 64 per cent of the population had no electricity at all. Greenstone asserted that 'the real threat to energy access is that energy is not treated as a private good, but as a right'.[40] This is the view of political economy, that access to energy is a private good, that has been opposed by the moral economy of the urban poor.

The first decade of the Rio process

The Rio convention accepted the reality, that the danger of global warming requires reduced fossil fuel consumption, and set out a strategy based on international agreements between states. It precluded binding commitments and deadlines. When some commitments were made, at Kyoto, they were so watered down as to be useless. The focus on achieving them via market mechanisms, and the obsession with commodifying, and trading, the right to emit greenhouse gases through e.g. the European Emissions Trading Scheme, firmly subordinated the wider human interest to the aim of economic growth.[41]

The historic failure at Rio and Kyoto to take serious action cannot be understood solely by looking at the negotiation process. For most governments, and the international financial organisations, these talks were of secondary or tertiary importance, compared to supporting the liberalisation initiatives in which they saw the way forward for economic policy. In most rich countries, even elementary forms of regulation that would have enabled reductions of fossil fuel consumption – e.g. monitoring emissions, phasing out fossil fuel production subsidies, regulating industries and transport, altering tax regimes – were either not applied at all, or applied too inconsistently to produce significant results. This was true even of the European governments that acknowledged the need for action to tackle global warming, and of the Democratic administration in the USA from 1993.

Two ideological trends took shape in the rich country elites that dominated the Rio process. The first, widespread in the US Republican party and the elites of Russia and other oil-producing countries, opposed regulatory action to deal with global warming on principle and embraced climate science denial. This denialism was abandoned by most big oil companies in the late 1990s, but persisted in political elites, merged with religious disavowal of the theory of evolution and assorted conspiracy theories, and became fertile ground for 'post-truth politics' in the 2010s.[42] This extremist ideology also exalted rich world people's right to excessive material consumption. As US delegates repeatedly insisted at the Rio conference: 'The American way of life is not up for negotiation.'[43]

The second ideological trend, represented by the US Democrats and European governments, accepted that global warming was a real threat and required action, but sought to reconcile that action with – and, in practice, subordinate it to – 'economic growth'. In many cases, this aim was understood in neoliberal terms, i.e. that unfettered free markets were the way to achieve this growth. This approach fitted with conservative attitudes to the potential for reducing rich-country energy demand, explanations of which, from engineers or environmentalists, were largely ignored.[44]

These trends clashed fiercely. Both of them influenced the policies adopted. The result was that the imperatives of capital accumulation trumped the need for collective state action articulated at Rio. This points to conclusions about the mechanisms of international governance, and the leading states that dominate them. Faced with the serious threats to future generations inherent in global warming, they failed to act effectively, and have continued to fail. The claims they make to represent the common interest of the whole of society cannot be accepted; they subordinated that interest to short-term requirements of the economically dominant parts of society.

A function of the Rio process was to create a discourse, legitimising the course of action taken. Global warming would be dealt with by market instruments, calibrated on the basis of climate science, and supervised via the intergovernmental agreements; society would be represented at the negotiations by the interaction of governments and NGOs. Ideas that global warming requires that the structure of the economy, of society be shaken up, were marginalised.[45] A preliminary conclusion is that to counter global warming, the issue must be taken out of the straitjacket to which the Rio process has confined it.

10
The 2000s: acceleration renewed

Some long-term fossil fuel consumption trends, that originated in the 1980s and 1990s and took shape fully in the 2000s, are outlined in the first section of this chapter. Further sections cover the rapid increase of consumption in 2000–2009; changes in the electricity sector; and energy policies and their failure to counteract consumption growth. The final section covers renewed consumption growth after the 2008–9 financial crisis, and assesses the beginnings of the transition away from fossil fuels.

Some long-term trends

Global fossil fuel consumption growth accelerated in the 2000s in the context of long-term trends that had taken shape from the 1980s. These included:

- *Fossil fuels' share of total commercial energy consumption fell, and then stabilised – but remained above 85 per cent.* (See Figure 8.) Between the mid-1970s and mid-1990s fossil fuels' share fell significantly, from 94 per cent to 87 per cent, due to investment in hydro and nuclear resources for electricity generation. In the mid-1990s, it stabilised at around 87 per cent. It rose to nearly 88 per cent in 2007, and then fell to 86 per cent, thanks to renewables, as well as hydro and nuclear. These were fossil fuels' shares of *commercial* energy consumption. If non-commercial biofuels are included, then fossil fuels' share of total energy consumption was the same in 2013 as a quarter of a century earlier, around 82 per cent.[1]
- *Coal became the fastest-growing fuel.* Since the 1980s, oil's share of total energy consumption remained, and remains, the largest, but has fallen. Gas's share has risen slightly. Coal's share has grown most significantly, and has accounted for the lion's share of consumption growth since the turn of the century. While coal supply to industry generally fell in rich countries, it rose elsewhere.
- *Fuel consumption grew most rapidly outside the rich world.* In the OECD, in the 1990s, consumption levelled off in some countries, although not the USA; in the 2000s, it fell in most countries – in some, quite substantially. Outside the rich world, consumption levels began to rise in the 1980s–1990s, but fell in the former Soviet countries in the 1990s.

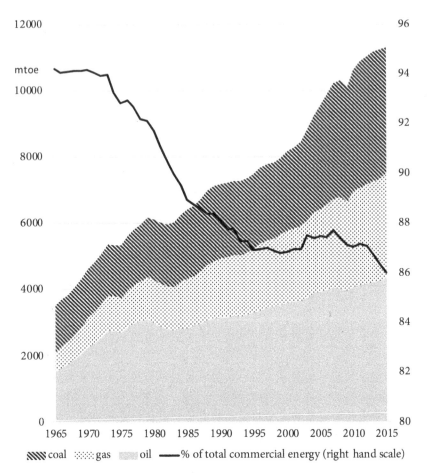

Figure 8 Global fossil fuel consumption, 1965–2015.
Amounts by fuel, and fossil energy as a proportion of total commercial energy consumption

Consumption growth accelerated in the 2000s, especially in China. (Figure 9 shows the trends in eight countries.) By the mid-2000s, China overtook the USA as the world's largest consumer of commercial energy, and consumption in non-OECD countries surpassed the level in the OECD – although their per capita consumption was still far, far lower. The shift was caused, most significantly, by the export of energy-intensive industrial capacity from the rich world to developing countries; consequently, rich-world industries' energy intensities, which had started falling in the 1980s as fuel prices rose, continued to fall.[2] Overall, between 1980 and 2015, industry's share of fossil fuel consumption in the OECD fell from 40 per cent to 31 per cent, and outside it rose from 28 per cent to 52 per cent. (See Table 15, p. 211.)

- *Electricity, heat and cars, as well as industry and the military, remained big fuel consumers.* The electricity and heat sectors globally remained the largest fossil fuel consumer. Oil for transport – in particular, private cars – remained a major source of consumption.
- *Household consumption, particularly of electricity, rose consistently.* In the rich countries, the amount of energy used per household rose consistently. In the 1960s, space heating had accounted for more than three-quarters of rich-world household energy use, and water heating was the second big item. But by the 2000s, electrical and electronic appliances had overtaken water heating.[3] In addition to this household consumption, in the 2000s the data centres that underpin the Internet became a big new source of electricity demand.
- *Increased fossil fuel consumption, particularly outside the rich world to supply goods to the rich world, was part of the overall trend for consumption of materials globally to rise.* A major study, using a newly established database of material flows, showed that, while some material consumption growth had been mitigated in 1970–90 by efficiency measures and innovation, there had been little improvement in efficiency since 1990, and that it had declined in the 2000s, mainly thanks to industrialisation and urbanisation outside the rich world. Another study challenged claims that economic growth was 'decoupling' from resource use. Decoupling was 'smaller than reported or even non-existent'; rich countries 'tend to reduce their domestic portion of materials extraction through international trade, whereas the overall mass of material consumption generally increases'.[4]

Global boom and bust, 2000–2009

The 1997 Asian economic crisis, and the 1998 Russian financial crisis, were followed in 2000–2001 by the bursting of the US 'dot com bubble', which wiped $8.5 trillion off firms' value and triggered a brief recession in the world's largest economy. Thereafter, swollen debt burdens and an expanding current-account deficit allowed the USA to become the engine of a new boom, in which industrial output growth in China and eastern Asia played a central role.[5] For the next decade, global fossil fuel consumption grew at a faster rate than at any time in history. In 2001–2010, global commercial energy consumption rose 28.2 per cent, compared to a 14.5 per cent increase over the previous decade. Almost the entire increase came outside the OECD.[6] The fossil fuel consumption boom was interrupted by the 2008–9 economic crisis, which caused consumption to fall for a year before resuming its upward trajectory.

Coal consumption rose the fastest, by 50 per cent over the decade to 2010. China, where consumption more than doubled, accounted for four-fifths of that

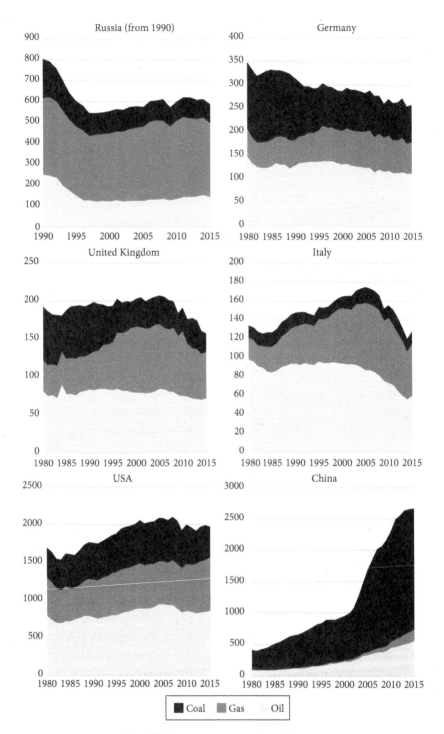

Figure 9 Commercial fossil fuel consumption in selected countries, 1980–2015

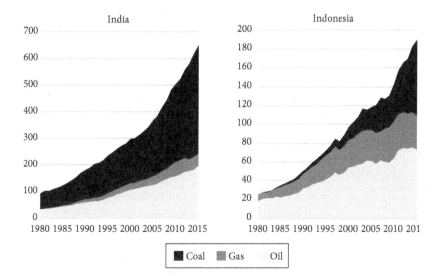

increase. China's share of world coal use rose from 30.5 per cent to 48 per cent. (See Table 10.) Much of the rest of the increase was in India, Indonesia and other east Asian countries. In parts of the rich world – including the USA, where gas began to be produced from shale rocks and other unconventional sources – gas displaced some coal from the electricity sector. But the expansion of coal-fired power in Japan (even before the Fukushima nuclear disaster of 2011), parts of Eastern Europe, and across the developing world, more than made up for that.

Investment in coal production leapt up internationally, from $30 billion in 2000 to around $75 billion in 2013. China's coal production galloped upwards: at the height of the boom, extra capacity of 3 million tonnes/year (roughly two or three large modern mines) was coming on each *week*, but production still could not keep up with demand. Mineworkers paid a high price, as they have so often: at least 23,418 of them died in accidents in China in 2001–8. China's coal exports decreased, until in 2009 it became a net importer; this was one cause of a spike in world coal prices in 2008. The boom also led to a new round of investment in electricity generation – which would result, when the economic crisis of 2008–9 came, in a new bout of overcapacity.[7]

Table 10 China's role in the 2000s

Commercial energy consumption, mtoe		Global	China	*China as % of world*
2001	Coal	2416.5	737.5	*30.5%*
	All fossil fuels	8261	992.2	*12.0%*
2010	Coal	3634.3	1743.4	*48.0%*
	All fossil fuels	10600.9	2291.3	*21.6%*

Source: *BP Statistical Review*

Much of the developing countries' coal use was for new, often energy-intensive, industrial production. Between 1999 and 2010, the share of the world output of steel – the number one industrial energy consumer – by the BRICS countries (Brazil, Russia, India, China and South Africa), rose from 28 per cent to 58 per cent. And by 2005, non-OECD countries were producing 78 per cent of the world's cement, 50 per cent of aluminium, 42 per cent of steel and 57 per cent of nitrogen-based fertilisers. Although global industry's energy intensity kept falling – the IEA calculated in 2013 that it was 12 per cent lower than in 2000, mainly thanks to older machinery and equipment being replaced – total fuel consumption by industry kept growing.[8]

Globally, consumption of oil and oil products, mostly for transport, continued to rise in the decade to 2010. Oil demand, like coal demand, grew fastest in east Asia and some other developing countries. By 2012, China, India, Japan, Singapore, South Korea and Taiwan were buying 45 per cent of global oil exports. A corollary of rising demand was an unprecedented boom in world crude oil prices. From their 1998 nadir of $12/barrel, they doubled by 2000, fell back briefly, and then climbed, surpassing $50/barrel in 2005 and $100/barrel in 2008. After peaking above $140/barrel, they crashed to around $30 in 2009, before rising again to a range of $80–110/barrel in 2010-14. Oil demand shocks, driven by global economic expansion, accounted for the bulk of the price increases, economists' studies indicated; financial speculation was probably more effect than cause. Clearly there had been no sudden rebalancing of power between producer and consumer countries, as in 1973, and no supply shocks comparable to 1979.[9]

For final consumers, petrol prices increased – but not sufficiently to change rich country drivers' behaviour substantially. In 2008, when oil was more expensive than ever, economists who compared 150 studies of price elasticities (i.e. the response by purchasers of petrol to changes in its price) concluded that end consumers are not very price sensitive, and even less so in the USA, Canada and Australia than in other countries.[10]

Rising oil prices caused severe financial problems for some developing countries, as they had done in the 1970s. By 2008, some international food commodity prices had spiked upwards, along with oil prices and partly due to their effect on agriculture, and this exacerbated the problem. Governments faced calls to control prices of diesel, kerosene and other oil products on which poorer households heavily relied. Very often they provided subsidies, by regulating retail prices; sometimes, in the face of social opposition, they dropped plans to let prices rise. Across southern Africa, there were violent protests over fuel prices in 2007–8. In Mauritania and Cameroon, price increases planned in 2008 were scrapped. In Namibia, the government forestalled protests by requiring energy companies to discount prices, and compensating them from the state's coffers.[11]

China's boom

China played a prominent part in the fossil fuel consumption boom, mainly due to its exceptional role as a producer of energy-intensive industrial goods for the world economy. Chinese government action was a factor: in the late 1990s, after two decades of market reform, it adopted a development strategy based on the rapid industrialisation of eastern coastal areas to serve export markets. While keeping big industries in state hands, it decentralised decision-making, giving a greater role to local authorities, and allowed private business to thrive in service sectors. Rich-world governments and businesses seized the opportunity to outsource heavy industrial tasks to China, which had the world's largest supply of cheap labour.[12] The result, a group of China energy specialists wrote, was that after WTO accession in 2001,

> China entered a highly energy-intensive growth phase fuelled by the inter-linked processes of industrialisation and urbanisation. This recent surge of energy consumption was driven largely by structural change in the economy, specifically, a steady shift towards heavy industrial production of materials such as steel, cement and chemicals. Although these energy-intensive industrial sectors have continued to improve their overall efficiency, production volume has grown more quickly and concentrated China's energy consumption and economic growth in the industrial sector.[13]

Between 2000 and 2007, production of steel, pig iron and aluminium more than doubled; cement production rose sixfold; and fertiliser production fivefold. China became by far the world's largest producer of steel, cement and aluminium. Here was a giant machine for turning fossil fuels into commodities: industries used three-quarters of China's primary energy supply, compared to about one-third in the rich countries. The machine was substantially less efficient than those elsewhere. China used 50 per cent more energy inputs to make steel and aluminium, and 40–46 per cent more energy inputs to make ethylene and ammonia, than world best practice. The result was a dramatic turnaround in the energy intensity of the economy. From 1980 to 2002, China's energy demand grew at half the rate of its GDP, so energy intensity fell; from 2002 onwards, its energy intensity galloped up by 5 per cent/year.[14]

As greenhouse gas emissions by China and other big exporters went through the roof, the issue of where the responsibility for those emissions lay – with exporting countries such as China, or with those mainly rich countries that bought the exports – became a focus of the international climate negotiations. Researchers responded by measuring consumption-based emissions, a procedure that attributed greenhouse gases emitted to those who consumed the exported goods, rather than those who produced them. (See pp. 205–6.)

It was estimated that, in 2002–8, almost half (48 per cent) of China's emissions were generated by industries producing goods for foreign consumption. The USA (34 per cent), Europe (26 per cent) and Japan (16 per cent) were the top consumers of emissions embodied in Chinese exports.[15]

China's export-led industrial boom was underpinned by a huge surge of urbanisation, which substantially increased fossil fuel consumption for construction of houses, roads and other infrastructure. It stimulated a sharp increase in household consumption for heating, cooking and electrical appliances. Between 1990 and 2007, the number of urban residents rose by 290 million, bringing the total urbanisation rate to 45 per cent. The total stock of urban housing space rose sixfold between 1990 and 2005; and urban households were moving from coal to electricity and natural gas. Stoves, boilers, refrigerators, TVs and clothes washers became widespread.[16] The boom wreathed China's cities in deadly air pollution.

The growth of a domestic car industry was integral to the Chinese government's vision of economic expansion. Private vehicle ownership was only made legal in 1984, and for another decade after that, manufacturers focused on truck production. In the late 1990s, the government sought to expand auto manufacture substantially, and made car ownership by each family a policy objective. Private vehicle ownership rose from 2 million in 1994 to 8 million in 2001 and 73 million by 2011.[17]

Electricity (2000–2010): liberalisation retreats, investment advances

The global electricity sector was in the early 2000s shaken by crises that discredited and pushed back liberalisation. There followed a renewed surge of demand, and investment. By 2010, a significant chunk of this investment, especially in Europe, was going to generation from renewables, raising hopes that the transition away from fossil fuels had begun.

Of the crises, the most dramatic – in 2000–2001 in California, the USA's most populous state – epitomised the extremes of the liberalisation drive, and the dangers to electricity infrastructure inherent in it. A surge in demand in 1997–2000, partly due to the growth of computer-related industries in California, put an unexpected strain on the system – although claims that there was any physical shortage of generation capacity are hotly contested. In any case, during 2000, amid fears of shortages, wholesale prices leapt up 13-fold, and some market prices 120-fold. Distribution companies buying from the wholesale market and selling to customers were being gouged for high prices that they could not pass on (due to retail market regulation). One of them, Pacific Gas & Electric, was bankrupted. Enron and other traders who profited handsomely were accused of creating artificial shortages at peak times. In 2001 the state authorities took a large measure of control of the sector, and ran up $8 billion of debt buying electricity and covering utilities' short positions. Customers not only paid four times more for electricity in 2000 than in 1999, but also suffered power cuts like those that are usual in developing countries.[18]

In the aftermath, the claim that 'insufficient [electricity] supply' was a fundamental cause of the problem was challenged by energy reform advocate Amory Lovins. He argued that California, having previously been a 'world leader in demand-side management', had in the mid-1990s adopted rules that rewarded firms that sold more electricity and penalised those that sold less. Debates also raged about how to apportion blame to traders who gamed the system, and regulators who patently failed in their responsibilities to keep wholesale prices 'just and reasonable'. Some of these arguments were settled by the bankruptcy of Enron in December 2001. Enron, which had started as a natural gas trading business, moved in the 1990s into electricity trading in the USA, and into investment in IPPs internationally. It took full advantage of its political connections (its CEO Kenneth Lay was a close friend of vice president Dick Cheney) and the creative talents of its accountants, Arthur Andersen, to run up extravagant bank debts and to build an elaborate pyramid scheme out of trading instruments. The collapse was the biggest in US corporate history. Suddenly, deregulation looked dangerous. Many US state governments reformed regulation, by, for instance, introducing price caps. In August 2003 the shakiness of market design was underlined by a power cut that left areas of the eastern USA temporarily without electricity.[19]

The reform of the Russian electricity system in 2006 was an indication of how liberalisation euphoria had subsided. Privatisation mania had been at its crudest in Russia in the 1990s: most oil companies had been sold at bargain-basement prices to the oligarchs (politically powerful businessmen) in the so-called 'sale of the century'. The electricity sell-off now struck a sharp contrast. Generation, transmission and distribution were separated-out and a closely regulated wholesale market set up. Retail tariffs remained fixed. Generating companies were sold, but with strict investment conditions; no risk-shedding long-term contracts were on offer as they had been to developing country IPPs. After the 2008–9 economic crisis, many generating assets were effectively renationalised – mostly by means of sale to state-controlled holding companies – after their private owners failed to renegotiate burdensome investment commitments that they had made on the basis of over-optimistic assumptions about economic growth and electricity demand.[20]

Globally, a new wave of electricity sector investment began, against the background of the 2002–8 boom and the retreat from deregulation. The annual total invested swelled from $290 billion in 2000 to almost $650 billion in 2012, the IEA estimated. To the extent that the state had resumed a directing role, particularly in the now-expanded European Union, a steadily growing share of these funds went into renewable generation. Global investment in renewables (mostly in generation) rose from $52 billion in 2004 to $243 billion in 2010. Germany, Spain, Denmark and other European countries provided subsidies to renewables, in the form of guaranteed payments for electricity supplied to the national grid. Some US states required companies to raise the proportion of electricity produced from renewables; those that offered support to household solar panels provoked coordinated opposition from utility companies. Such subsidies remained far smaller than those for fossil fuels, but enabled some countries to consolidate renewable generation on a significant scale.[21]

During the 2000s, electricity demand for information technologies grew at dizzying speed. The digital revolution had been changing demand patterns since the 1990s, with personal computers, mobile phones and CD players replacing older gadgets, but did not produce electricity savings for which conservationists hoped. By the early 2010s, US households were using more than twice the amount of electricity they had in the early 1980s; electrical and electronic appliances, including air conditioners, accounted for 40 per cent of electricity consumption in buildings; 99.6 per cent of households had a TV, and one third had more than one. Standby mode became a burden: in the USA in the late 1990s, up to 40 per cent of the electricity used by electronic equipment was in standby mode, and an empty US household was on average leaking 50 Watts of electricity, 49 W of which was needless.[22]

After the turn of the century, the new generations of Internet and mobile technology (the 'cloud') that required continuous data provision produced a new type of demand. Estimates of electricity demand, compiled in the 2010s in the teeth of ferocious secrecy by most big Internet companies, range from 263 bn kwh/year (including 76 bn kwh in the USA) for data centres, to 372 bn kwh/year for data centres and telecoms infrastructure. Consumption by the whole 'cloud', including manufacture and use of devices, has been estimated at 623 bn kwh/year – more than total consumption for all purposes by India, the fifth-biggest country by electricity use. Consumption by the entire information and communication system has been estimated at 1500 bn kwh/year. Much electricity is used egregiously for multiple levels of backup for Internet services, a detailed investigation by McKinsey and co., the business consultants, found. 'Most data centres, by design, consumed vast amounts of energy in an incongruously wasteful manner', the *New York Times* showed in 2012; the centres were using 6–12 per cent of the power to perform computations; the rest was essentially 'to keep servers idling and ready in case of a surge in activity that could slow or crash their operations'; the availability of such extra capacity encouraged wasteful habits among computer users.[23]

The failure of energy policy to deal with global warming, 2000–2017

The acceleration of global fossil fuel consumption growth in 2002–8 was, on the face of it, proof enough that the international climate change negotiations had failed. The negotiators accepted that fossil fuel use is the primary cause of greenhouse gas emissions and so needed to be curbed, and claimed to have acted accordingly – but fossil fuel use rose still more rapidly. Despite this, the vast bulk of commentary continues to portray the process as the only, or the most significant, vehicle for the transition away from fossil fuels, or, at least, accepts that such a portrayal deserves to be taken seriously. How do we square the circle?

The 1992 Rio treaty was essentially a statement of good intentions. The 1997 Kyoto agreement was the first, and so far the only, agreement that committed countries to reducing emissions of CO_2 and other greenhouse gases by specific amounts. It came into force in February 2005. The treaty required 36 (later 43) developed nations and former 'socialist' countries to reduce greenhouse gas emissions by an average of 5 per cent between 2008 and 2012. The target was easily met, with emissions by those countries collectively about 16 per cent lower in 2012 than in 1990. But emission reductions were counted so as to include (i) the reduction caused by the 1990s slump in post-Soviet countries, and (ii) the effect of countries buying carbon credits, or swapping them with developing countries via a 'clean development mechanism', to 'meet' their targets. Moreover, rich countries' emissions reductions were mainly due not to the agreement, but to the shifting of carbon-intensive industrial processes to developing countries, and to the recession of 2008–10.[24]

The Kyoto agreement was designed to be replaced by a similar deal, covering the period after 2012, which was due to be finalised at the UN Climate Change conference in December 2009 in Copenhagen. A month beforehand, hopes that the USA would participate were raised by the election of a Democratic president, Barack Obama. In the event the conference failed to agree on any actions or binding targets. The next major conference, held in Paris in December 2015, formally ended efforts to set international targets for reducing emissions. Agreement was reached between 196 countries to constrain global warming – keeping average temperatures 'well below' 2°C higher than pre-industrial temperatures, and 'endeavour[ing]' to keep them below 1.5°C higher – by means of Nationally Determined Contributions (targets set voluntarily by national governments). These pledges amount to a collective abandonment of the declared aim. Climatologists estimated that, if all the pledges were kept, global average temperature will rise by 2.7°C, as opposed to the 1.5–2°C targets (and by 3.6°C if policies are unchanged). The economist Jamie Morgan calculated that, if all the pledges were kept, the nations that made them would collectively miss the IPCC's 'least cost' scenario (i.e. an estimate of the level to which emissions need to fall, if a transition away from fossil fuels is to be economically efficient) by 4.7–11 Gt CO_2 in 2025 and 11.1–21.7 Gt in 2030.[25]

It is tempting to dismiss the entire negotiation process as window dressing, designed to obscure governments' collective failure to tackle global warming

– but that is not convincing. Governments need to be seen to act on behalf of society as a whole; the Rio process reflects, in part, the pressure they have been subject to by broad social concern about global warming. For social movements urging robust measures, the Paris declarations may have helped create space for political, and even legal, action. But Paris also shone a harsh light on the failure of the process as a whole. The principal participants largely rejected strong regulation, and stuck stubbornly to the principle of reducing emissions by commodifying them (i.e. producing tradable rights to pollute). Most governments, most of the time, subordinated the talks to imperatives of 'economic growth' and associated policy priorities, including continued support for the fossil fuel industries and the car industry. The effect was to render the process ineffectual.

After Kyoto, the European Union emerged as the signatory that made the most plausible attempt to minimise greenhouse gas emissions. It did not do so quickly. Revisions to the EU's founding treaty in 1992, 1997 and 2001, in keeping with the neoliberal times, all focused on deepening the single European market. Only with the Lisbon treaty of 2007 was a serious attempt made at a unified energy policy that included emissions reduction. In the early 2000s, the EU's most notable contribution was to secure Russian ratification of the Kyoto treaty. The Russian government had followed its Soviet predecessor in vehemently opposing action to reduce emissions. But by making 1990 the date with which future emissions would be compared, negotiators were able to secure a free pass for Russia and other former Soviet countries, whose share of emissions reduction had been more than met as a result of the slump of the early 1990s. This, together with EU support for Russia's application to join the WTO, proved persuasive.[26]

In 2005 the EU set up the Emissions Trading System, which became the world's largest carbon market. EU member governments issued permits to companies allowing them to emit certain quantities of CO_2 and other greenhouse gases, which could be traded, with a view to forming a market price for emitting carbon. This, it was claimed, would encourage businesses to reduce emissions. The market has failed, by its own and any other standards. Too many permits were issued; corruption surrounding the award of permits was widespread; and, most importantly, the price of the permits, instead of going up – which would, at least theoretically, have acted as a lever for energy efficiency improvements – crashed repeatedly. In May 2006 it fell from €30/tonne of carbon to under €10/tonne. It climbed again to €20/tonne in early 2008 and collapsed again to €3.5/tonne in early 2013 – at a time when industry participants saw €100–200/tonne as the levels needed to effect substantial change. The European Parliament then *rejected* a proposal to withdraw some permits from the market to support the price. The IEA estimated that 8 per cent of global carbon emissions were subject to the (uselessly low) carbon price, while 15 per cent of emissions received an incentive of on average $110/tonne, in the form of fossil fuel subsidies.[27]

In 2007 the European Commission added to the carbon market policy a set of targets for member states aimed at reducing emissions to 30 per cent below the 1990 level by 2020. This has underpinned support for renewable electricity generation – although that was not uniform, and some member states scrapped such measures after the 2008–9 economic crisis. In addition, the Commission has used regulatory measures to support the phasing-out of coal-fired electricity generation (the Large Combustion Plant Directive of 2001 and the Industrial Emissions Directive of 2011), aimed at reducing health-threatening pollutants (sulphur dioxide and nitrogen oxides).[28]

The two other largest participants in the climate talks, the USA and China, also made progress on greenhouse gas emissions reduction, especially when it dovetailed with air pollution regulation. In the USA, federal government action stalled entirely under president George Bush (2001–9); Obama, from 2009, pushed through fuel efficiency standards for cars that Bush had blocked, and overturned obstacles to state-level regulation (e.g. in California) imposed by Bush administration officials. In China, robust energy efficiency policies were written into the 11th Five Year Plan (2006–10) and further strengthened in the 12th (2011–15) and 13th (2016–20) Plans. Measures were adopted to reduce coal-fired electricity generation, to move it to less populated western regions, and substantial investment made in wind power. The main political impetus for constraining coal in China was to reduce air pollution, rather than global warming; this has also become pressing in India and other Asian countries.[29]

The collective timidity of government measures aimed directly at tackling global warming can be better understood in the context of a wider range of policies, including those on the regulation of fossil-fuel-consuming activities per se; on support for energy-conserving technologies; and on fossil fuel subsidies.

A good example of weak regulation is that of fuel efficiency standards for cars. In the USA, fuel efficiency standards, which had been frozen since the 1980s, were tightened in 2007, but any positive effect was cancelled out by the trend towards replacing cars with SUVs. (See p. 128.) Researchers pointed out that if US drivers used vehicles that weighed the 1987 average, they could consume 24 per cent less fuel; if the 2008 Europe average, 30 per cent less fuel.[30] Managers at vehicle manufacturers made concerted efforts to evade regulation by falsifying test results, both for various emissions and for fuel efficiency. This led in 2016 to a scandal at Volkswagen and other companies. Overall, cars' fuel economy improved by only 2 per cent in the decade 2006–15, while engineers expressed confidence that cars *could* be produced that were up to ten times as fuel-efficient as existing ones.[31]

In respect of state regulation of household energy use, Japan was the exception that proved the rule. Its Top Runner programme, launched in 1998, set mandatory efficiency improvement standards for manufacturers of household electrical goods. By 2005, new air conditioning systems used 68 per cent less electricity than those sold ten years previously; computers, 81 per cent; refrigerators, 55 per cent; and so on. These figures are many times greater than those for countries with less or no regulation. US regulators, by contrast, showed how to make matters worse over time with air conditioning, one of the most energy-intensive household technologies. The standard indoor temperature aimed at in summer had by the twenty-first century fallen to 21–22°C, from 25.5°C in 1925; many US builders aimed as low as 19°C for public spaces (e.g. shopping malls).[32]

Government attitudes to new energy-conserving technologies have been lukewarm. One example is that of zero-energy buildings. Progress has been slow, the IEA concluded, because low energy prices act as a disincentive to businesses scaling up the technologies that are available, and lower cost technologies are not yet developed. A second example is that of distributed generation in electricity. (See pp. 36–7.) Many technological obstacles have long ago been overcome, but electricity companies are unsure how to proceed without cutting into their revenues. Governments have done little.[33]

The patchy support for technologies that could hasten the transition away from fossil fuels strikes a sharp contrast with the steadfast backing for fossil fuel production and consumption, via subsidies. Overall, global fossil fuel subsidies swelled enormously in the 2000s: the total was by the 2010s estimated at more than $400 billion/year. The largest slice, particularly in 2003–8, were consumer subsidies for petroleum products, which governments paid as soaring oil prices produced fears of economic disruption and social unrest: these rose from $57 billion in 2003 to $519 billion in 2008, and fell back to $136 billion in 2009 and $240 billion in 2010, IEA researchers calculated. There was also strong government support for fossil fuel producers, particularly in rich countries.[34]

In 2010, the OECD significantly boosted subsidies research, by drawing up an inventory of fossil fuel subsidies in its member states. OECD researchers tracked more than 550 government measures, of which two-thirds were tax breaks and 70 per cent supported oil and oil products. In 2005–11 these totalled $55–90 billion/year. By 2015 the inventory had been expanded to include the largest non-OECD economies and recorded $160–200 billion/year of measures in 2010–14, with indications of a downward trend as oil prices fell in 2013–14. Russia, the USA, Australia, Germany and the UK were the worst culprits. Among the particularly egregious subsidies highlighted were the cost-depletion tax break for US oil producers ($1.19 billion in 2011) and the accelerated capital cost allowance for Canadian oil sands production ($300 million/year in 2007–11).[35] To these numbers may be added the effect of the non-taxation of aviation fuel, dating back to an international agreement of 1944. (See p. 23.) Shipping fuel has also avoided tax, due to international custom and practice developed in recent decades. The implied tax break is at least $38 billion/year, IMF staff researchers reckoned.[36]

The OECD's research was built on by NGOs. A project focused on state support for oil and gas exploration found, in G20 nations, $16.3–$22.6 billion of direct state spending e.g. on seismic surveys and tax breaks for drilling equipment purchases; $48.7–$49.4 billion investment by state companies; and $14.7–$15.9 billion in financing e.g. by multilateral development banks. Another project tracked the $9 billion/year in public finance for coal producers approved in 2007–14, particularly by rich countries' export credit agencies. (These sums are gross commitments; the portion defined as a subsidy would be calculated differently.) Perhaps the best measure of political intentions, though, was a report on the implementation of pledges made by leaders of G20 nations in 2009, in a blaze of publicity, to phase out fossil fuel subsidies. Governments had, over the two-year period monitored, been 'changing their definitions, not their subsidy policies'; they had failed to keep their own promises to track and report on subsidies; the standard of governments' reporting was 'well below any reasonable minimum needed for real reforms to take hold'; and, surprise surprise, 'no subsidies have actually been eliminated as a result of the G20 commitment'.[37]

The 2010s: a time of extremes

Global fossil fuel consumption bounced back from the 2008–9 crisis very rapidly. The grand total fell by 1.87 per cent in 2009, but rose by 4.7 per cent in 2010 and 2.35 per cent in 2011, keeping up with world GDP growth. From 2012, though, global fossil fuel consumption growth slowed down each year, falling below 1 per cent per year, while world GDP growth stayed above 2 per cent.[38] In 2013–14, with fossil fuel consumption growing only slightly, coal use in the OECD countries falling, and gas outpacing the other fuels, CO_2 emissions from burning fossil fuels were constant. Advocates of the UNFCCC process declared that GDP growth and emissions are 'decoupling' – although two or three years' data hardly confirms a trend – and that the world was now on the way to 'green growth'.[39] Actually, while a shift towards electricity generation from renewables had started in some countries, fossil fuels remained completely dominant in electricity and other energy-consuming systems (transport, industry and urban infrastructure) up to the mid 2010s. Technologies on which to base a transition away from fossil fuels became more widely available, but the transition has hardly begun.

Coal consumption growth slowed in the mid-2010s. 'Green growth' enthusiasts have equated this to the final decline of coal, but the future extent of the decline remains unclear. In the USA, the 'shale gas' boom meant that cheap gas displaced coal from electricity generation; oil companies looked to gas as the best competitor for renewables. In Europe, although coal use increased in some countries, the growth of renewables, sluggish economic activity and regulatory action combined against coal. In China, coal use stopped growing – and may even have fallen, although there are question marks over statistics – thanks to slower economic expansion, energy efficiency improvements in industry and air pollution policies.[40] Meanwhile, coal consumption surged in other developing countries. In southeast Asia, 25 GW of coal-fired electricity generation capacity was added in 2010–15; there and in India, cheap coal diverted from the Chinese market was swallowed up at low prices. Coal consumption rose in Turkey and conflict-torn Ukraine, thanks to policies aimed at minimising dependence on imported Russian gas.[41]

In terms of sheer volume, this thirst for coal did not make up for lower demand in China in the USA. Doubts about future coal demand in China and India cast a shadow over the fuel's future. In 2015, there was over-capacity in coal production and coal-fired electricity generation, making for a 'perfect storm' in investment, the IEA reported. Construction of 68 GW of coal-fired electricity generation capacity was frozen at more than 100 sites in China and India, and in 2016, worldwide, pre-construction activity fell by half and construction starts by more than half.[42]

The clearest indication that a transition away from fossil fuels has begun in some countries, and of its limits, is the surge of investment in renewable electricity generation. Over the decade 2006–15, investment in electricity generated from fossil fuels was constant at $100–$130 billion/year, while investment in generation from renewables (mostly wind and solar power), which had been $60 billion in 2000, rose from $120 billion in 2006 to $250–300 billion/year in 2011–15. Renewables were by 2015 accounting for about half of new capacity additions; its costs kept falling. Denmark, Spain and Germany were generating a substantial proportion of their electricity from wind and solar (51 per cent, 22 per cent and 18 per cent respectively, compared to 7 per cent across the OECD as a whole). But the battle between renewables and fossil fuels in the electricity sector remains one of David against Goliath. Between 1990 and 2015, renewables' share of electricity generation worldwide rose from 1 per cent to 5 per cent, while hydro power's fell from 18 per cent to 16 per cent and fossil fuels' share rose from 63 per cent to 68 per cent.[43] In 2014–15 investment in extraction of oil, gas and coal ($580 billion in 2015), and their transportation to customers ($320 billion in 2015), continued to dwarf all other energy investment flows.[44] And outside the electricity sector, i.e. with respect to transport, industry, urban infrastructure and energy efficiency, the evidence, at very best, points to the first tiptoe steps away from fossil-fuel-dominated systems.

In transport, trends have left researchers divided on interpretation. In OECD countries, oil product consumption for transport fell slightly after the 2008–9 crisis, while car ownership rates continued to rise, to one vehicle for every two people (OECD average) and 1.5 vehicles for every two people (US average), by 2012. In some cities public transit, cycling and walking have been revived.

The transport specialists Peter Newman and Jeffrey Kenworthy reported that growth of car use 'appears to have plateaued and is now beginning to decline' in the rich world, and optimistically forecast 'the end of auto dependence', not only in the OECD but beyond it. Urban transport researchers at the UN have a more cautious view. In 2013 they found that the share of travel by public transport, as opposed to private cars, had 'decreased or stagnated in most developing country cities'. The parlous state of public transport pushes people back into private cars, they pointed out – or, in the case of the wealthiest residents in Sao Paulo, Brazil, into helicopters to avoid traffic. The sort of public transport revolution needed to cut fuel consumption significantly is still in front.[45]

To sum up, it is unclear at the time of writing whether the shift away from coal in electricity generation will consolidate as a trend; whether and how renewables can leap from occupying a corner of a centralised, fossil-fuel-fired electricity system to a dominant position; and how distributed generation and smart grid technologies can overcome the obstacle of the profit motive. It is unclear whether and how new technologies, such as electric cars, can beat the inertia of existing social and economic systems. Deep-seated changes in the way that industry, urban infrastructure and households use fossil fuels and energy-intensive technologies may be even further ahead.

Part III

Reflections

11
Interpretations and ideologies

The view offered in this book – that fossil fuels are consumed by and through technological systems, which are in turn situated in social and economic systems – rests on a set of interpretive starting points that themselves have a history. The ideas and assumptions most relevant to the book's theme are discussed in this chapter. First, about the way that technological, social and economic systems interact with natural systems, and what it means to do so sustainably. Second, about the role that consumption and consumerism play in the economy. Third, about population and neo-Malthusian approaches to its role in consumption. Fourth, about economic growth and its relationship to fossil fuel consumption.

Social systems, natural systems and sustainability

Efforts to understand and theorise society's interactions with nature go back thousands of years. In early nineteenth century Europe, the rapid expansion of industry, and the coal burning that went with it, produced not only widespread social concern about air pollution but also 'acute awareness' among philosophers and natural scientists of the importance of society-nature interactions, as Christophe Bonneuil and Jean-Baptiste Fressoz have underlined. By the mid-twentieth century, though, these interactions had largely been pushed out of the subject matter of the social sciences. Economics, in particular, was culpable, in excluding flows of natural resources from its analyses of production and consumption.[1] During the post-war boom, it was ecologists (including Rachel Carson, Barry Commoner, and Murray Bookchin in English-speaking culture) who noted the 'great acceleration' of human impact on the environment.[2] Their work influenced the emergence of environmentalism as a modern social movement in the 1960s, and its entry into the political mainstream in the USA and Europe in the 1970s.

The *Limits to Growth* report (1972) put the physical constraints on society's use of natural resources on the agenda of political elites. In the 1980s came the concept of sustainability, defined by the UN Brundtland commission report (1987) as humanity's ability to meet its present needs without compromising future generations' ability to do so without putting excessive pressure on the available natural resources. Sustainable development, the report stated, implied the existence of natural limits on human activity, 'not absolute limits', but limits 'imposed by the present state of technology and social organisation'.[3] But much sustainability research thereafter concentrated more on measuring the quantitative impacts of human activity on nature – for example, limits to the availability of fossil fuels, agricultural land, minerals, fresh water and other natural resources – than on the relative character given to those limits by social and technological factors.

'Sustainability' soon took on as many contested meanings as 'democracy' or 'justice'. In finance and business – and specifically in the international financial institutions – it became a cosmetic checklist that made little difference to economic activity. Economists found ways to reconcile sustainability with economic growth: Jeffrey Sachs declared sustainable development a 'normative approach', seeking 'a good society' that is 'economically prosperous [...] socially inclusive, environmentally sustainable and well governed'. Others sought to quantify the resources provided by nature to society in economic terms, as 'ecosystem services' – or, as the International Union for the Conservation of Nature put it, 'the biggest company on earth'. This approach was reflected in the 2000s in the establishment of carbon markets. Mainstream economic thought frequently linked such ideas with technocratic solutions, up to and including geoengineering to fix climate change; these were opposed by those urging deliberate constraints on growth, such as the authors of the *Limits to Growth* report.[4]

The foregrounding of sustainability in the 1980s, and the clarification of fossil fuels' causal role in global warming by the 1990s, provided conditions for new research of the way that energy, and materials, move through the economy. Holistic analyses of the flow of materials and energy through both natural and social systems were now developed by researchers in fields such as environmental sociology and environmental systems analysis. (See pp. 204–5.) I hope that this book complements that work.

The role of consumption, and consumerism, in the economy

Consumption, along with production and exchange, was studied by the social theorists of the late eighteenth and early nineteenth centuries, who sought to explain the human relationships forming as a result of capitalism's rise to dominance. The new discipline of political economy usually assumed that all people had a natural desire to accumulate wealth, and that everything of value in the economy's functioning was to be found in the final demand for goods and services. Adam Smith, one of the founders of political economy, wrote: 'Consumption is the sole end and purpose of all production; and the interest of the producer ought to be attended to only so far as it may be necessary for promoting that of the consumer.' Smith famously argued that producers of goods bring them to market to exchange them, not out of care for the consumer, but out of self-interest. From this, Smith's twentieth and twenty-first-century followers developed crude theories that markets, driven by self-interest, are to be welcomed as the natural order of things.[5]

Karl Marx challenged the political economists' assumption that market relations driven by self-interest were natural. This, he believed, was exactly what needed to be questioned. Accepting Smith's idea that commodities comprised both use value and exchange value, he argued that they manifested a contradiction inherent in the labour that produced them: it was both 'concrete labour' that produced the particular thing, and 'abstract labour' that produced the commodity's exchange value. This exchange value was appropriated by its owner, the capitalist. Marx saw these exploitative relationships between people in the context of human-nature interactions. In his eyes, labour – by which humans take from nature their means of subsistence and the basis of their culture – had, through history, become an unnatural activity. People who labour were 'alienated', from their natural environment, from each other, and – crucially, for understanding economic relationships – from the products of their labour. These products are appropriated by those with wealth and power, and become the basis for the constant expansion of that wealth and power. Capital, Marx famously proposed, is a social relation. What drives the economy is *not* primarily the satisfaction of consumer demand, and the self-interest of those who serve it, but the constant need to renew and expand the accumulation of capital.[6]

Marx's writings on consumption focused on the distinction between luxury consumption, by those with wealth and power, and the subsistence-level consumption by working people.

Consumption above subsistence level took shape as a mass social phenomenon, beyond society's elites, only from the late nineteenth century in rich countries. Early twentieth century social theorists such as Max Weber and Thorstein Veblen critiqued both consumption and the values and ideologies that accompanied it, that later came to be called consumerism. Twentieth century socialist analysts of consumption built on Marx's critique of the Smithian assumption that the economy's purpose was to satisfy demand, and contextualised consumerism in hierarchical social relations. Lewis Mumford described 'a sharp shift in interest from life values to pecuniary values'. R. H. Tawney criticised the 'acquisitive society' that promised to the strong 'unfettered freedom for the exercise of their [property-based] strength'.[7]

The boom of the 1950s brought new forms of consumerism – and a blistering attack on economists' 'conventional wisdom' about it from within their own ranks. The Keynesian J. K. Galbraith lambasted the theory of consumer demand (i.e. that it caused, and justified, the focus on output growth), on the grounds, first, that it makes no distinction between the satisfaction of physical needs (i.e. subsistence) and of 'psychologically grounded desires', and, second, that it promotes the fiction that 'wants originate in the personality of the consumer'. Consumers' wants, he argued, are 'contrived by the process of production'; 'the urge to consume is fathered by the value system which emphasises the ability of the society to produce'; advertising, which clearly cannot be reconciled with the notion of independently determined desires, creates desires; wants are 'dependent on production'.[8]

From the 1970s, sociologists and anthropologists pushed back against the idea of the manipulated consumer (e.g. Galbraith's assertion that wants are 'contrived'). The theory of 'consumer sovereignty' proposed that rich-world consumers, far from being helpless victims of advertising, used consumption to shape their lives. Such research produced important insights, although these were often devalued by leaving unclear the constraints on consumers' agency imposed by social realities, e.g. working people's dependence on wage labour. Other research illuminated the social construction of ideas about people's 'needs' (e.g. the passage from 'I *need* food' to 'I *need* to upgrade to the model with a more powerful engine'), and took apart ideologies of 'scarcity', which sought natural causes for scarcities bound up with social and economic systems.[9] Such approaches allow us to analyse consumerism as a social phenomenon used by, and reinforced by, political and corporate elites, without viewing consumers as stupid dupes of advertising.

In the 1980s, the American Marxist Allan Schnaiberg attempted to clarify theoretically the position of consumption – and, specifically, fossil fuel consumption – in modern capitalist economies. He argued against 'crude environmental analyses' that take 'total societal production and divide by the population, in order to achieve a "consumption per capita" figure', on the grounds that (i) 'a substantial share of production goes into other forms of production, never reaching the consumer directly' (both traditional producer goods such as machinery, but also commodities that support production and marketing activities, of which many are energy-intensive, such as transport); and (ii) because public service provision meant that some goods and services are provided without consumers' discretion. Schnaiberg argued not that consumer preference was irrelevant, but that constraints on it should be defined. These included *direct manipulation* (e.g. advertising), *preference constraints* (effects of income, the mismatch of available products and demand, problems of retrospective demand) and *the provision of public goods*.[10]

In the 2000s, researchers on household energy consumption built on Schnaiberg's work, and Daniel Spreng's net energy analysis, in pointing up the distinction between *discretionary* and *non-discretionary* energy consumption. (See p. 46 and p. 205.) I have endeavoured to bear this in mind in writing this book.

Population and consumption

Disputes about population growth as a cause of rising fossil fuel consumption, like broader issues of consumption's place in the economy, have their roots in early nineteenth century political economy. Adam Smith, and most social theorists, assumed that population growth would expand human productive capacity, and therefore prosperity. Thomas Malthus, in his *Essay on the Principle of Population* (1798), proposed a darker view: that an expanding population would need to consume more than the available natural resources would be able to provide. Population would grow geometrically, but the means of subsistence could grow only arithmetically. Malthus's notorious political conclusion was that relief payments to the poor should be stopped, since they obstructed checks on population growth such as poverty, early death and limits on family size. For this he was denounced by Marx, who saw it as evidence of the inhumanity of a capitalist ideologue.[11]

Malthus's assumptions have repeatedly been proved wrong – since he took insufficient account e.g. of the ability of technology to increase agricultural yield and supply a larger population from the same area of land – but the idea that there are absolute natural limits to population growth has remained influential, including in some strands of environmentalism.

During the 1950s and 1960s, in international institutions, in academia, and in some developing-country governments, the belief that population growth was potentially negative for the economy grew stronger. In some developing countries, it led to (usually unsuccessful) attempts at birth control. Neo-Malthusianism arrived with *The Population Bomb* (1968), a restatement of the case for compulsive population control, by the American biologist Paul Ehrlich. Another prominent population control advocate, Garrett Hardin, argued that humans' inherent selfishness excluded the possibility of sharing resources; he used an essentially racist analogy of a 'lifeboat' (from which 'we' must unfortunately exclude the poor and helpless) for society.[12]

The case against populationism was made by, among others, the biologist Barry Commoner, who argued that technological change, and the new types of production made possible by it, were the primary cause of society's impact on the environment, more significant than either population growth or rising living standards. Ehrlich was also challenged, notably, by Julian Simons, a mainstream economist who argued that he had taken no account of the potential for technological change to alter the way that natural resources were used. (See pp. 201–3.)

The focus by Ehrlich and others on population dovetailed with ideological dogma that focused development policies towards managing, if not controlling, population outside the rich world – and away from the effects of the rich countries' economic domination. At the Rio Earth Summit in 1992, disputes about the relative roles of population growth and rich-country consumption raged, not only between the government delegations, but also among NGOs seeking to influence the international talks. (See p. 139.) NGOs based in the Global South insisted that excessive consumption and lifestyle in the north had to be taken into account. Their northern counterparts argued that – even if they themselves believed that lifestyles should change – neither governments nor rich-world populations would countenance that. In NGO discussions parallel to the summit, one northern representative 'was worried [...] that the question of population growth in developing countries had not been adequately addressed', two participants recalled. 'The southern NGOs retorted [...] that the population problem was essentially in the North, where each child born is likely to consume much more, causing the emission of more greenhouse gases.'[13]

Research methods that give excessive weight to population, as an undifferentiated phenomenon, continued to be used in the work that fed into the IPCC's reports. (See pp. 202–3.) Such studies pushed into the background distinctions between different types of direct and indirect consumption, different types of consumers – and in particular, the distinctions between those in the rich world and those outside it – and the role of social, economic and technological factors.

In 2012, in an issue of *Nature*, the leading science journal, marking the twentieth anniversary of the Rio summit, Ehrlich, writing jointly with colleagues, asserted: 'each person added to the population will derive food and other resources from poorer sources, generally involving more energy and disproportionate environmental impact'.[14] The way that such arithmetic dominates public discourse is a tribute to the endurance of neo-Malthusian ideology. In the preceding chapters, I have offered a view of fossil fuel consumption with a contrasting emphasis: that it is social, economic and technological systems that consume resources, that individuals do so through those systems and that there is no direct, arithmetic correlation between their consumption and environmental impacts.

Economic growth ideology and alternatives

In the dominant public discourse, fossil fuel consumption growth is attributed to 'economic growth', alongside population growth. The IPCC's most recent Assessment Report, for example, states: 'Globally, economic and population growth continue to be the most important drivers of increases in CO_2 emissions from fossil fuel combustion.'[15] The report includes a mass of detail showing that the drivers are actually far more complex – but this is the summary, beyond which politicians and journalists usually do not read.

In the case of population growth, neither correlation with fossil fuel consumption growth, nor causation, can be concluded from the statistics. (See pp. 47–50.) Things are different with the relationship between economic growth, as measured by GDP, and fossil fuel consumption levels. Trends are often correlated. Globally, the temporary dips in energy consumption since 1950 – in 1974, 1979–80 and 2008–9 – all resulted from economic crises that halted growth. In the rich world until the 1980s, economic growth rates and fossil fuel consumption rates followed each other quite closely. Then, as energy-intensive industries were exported to the Global South and efficiency improvements made, rich countries' economic growth rates surpassed fuel consumption rates. In most of the Global South the reverse applied. Still, correlation does not indicate causation. Statistical comparisons of GDP and fuel consumption do not usually say much, and sometimes say nothing, about, for example, the structure of countries' economies, or geographical factors, that may be significant determinants of fuel consumption levels. They exclude analysis of energy flows through technological systems.

Why, then, is so much energy devoted to squashing analysis of fossil fuel consumption into such a badly suited framework? Because the ideological dogma of economic growth, as measured by GDP, pervades so much public discussion – about energy policy and the transition away from fossil fuels, as well as much else. Industrialisation, urbanisation and other trends associated with economic growth are assumed *a priori* to be positive. Pro-growth policies are assumed *a priori* to have been the only way to achieve these positive results; clearly negative aspects of growth – the effects of financial crashes, or the harmful results of liberalisation, as well as impacts on nature – are ignored or downplayed. So is the growing body of research by mainstream economists that has concluded, incontrovertibly, that growth has reinforced and deepened inequalities.[16]

Moreover, the equation of economic growth with prosperity assumes *a priori* that fundamental aspects of life under modern capitalism – such as systems of wage labour and domestic labour, the dependence of urban populations on industrial agriculture, and the forms of alienation inherent in these – are also sources of prosperity. The 'conventional wisdom' of growth-ism and consumerism reinforce each other, as Galbraith argued more than half a century ago.

A significant challenge to economic growth dogma was offered by proponents of ecological economics, notably E. F. Schumacher in the 1970s and Herman Daly in the 1990s.[17] From the starting point that economic growth is unsustainable because of its impact on natural resources, they went on to expose the vacuity of the claim that growth can be identified with human development, happiness or prosperity. Although the ecological economists have complained, justifiably, that their attack on orthodoxy has met with little by way of any theoretical defence, there has been an indirect response: efforts to reconcile sustainability principles with pro-growth economics, i.e. 'green growth'. This was the starting point for the Stern Review (2007), around which many mainstream discussions of energy policy have been oriented.[18]

From a socialist standpoint, the ecological economists' readiness to open up fundamental questions about ruptures in human society's relationship with the natural world, and the failure of mainstream economics to face them, seems to strike a sharp contrast to the assumptions some of them make, that the necessary changes can be made by existing power structures and within existing property relations. In some respect this is the mirror image of much twentieth century socialist ideology that believed that property relations and power structures had to change, only to continue with economic growth in a 'socialist' guise. This productivism was strongly influenced by the state socialists who ruled the Soviet Union and China, but was also widespread in labour movements. Attempts to rework socialist ideas to correct this blind spot were made by Andre Gorz in the 1980s. More recent eco-socialist and other radically minded writers have continued to address relevant issues, in dialogue with 'de-growth' advocates who, increasingly, see their aims in terms of profound social change.[19]

A vision of a future in which social change transforms not only property relations but also the labour process through which humans relate to nature is beginning to take shape. In my view, such a vision offers the most compelling alternative to the dogma of economic growth and the assumed inevitability of exploitation, inequality, and worse that it implies. Such a transformation would offer the best conditions for a transition away from fossil fuels. How and whether any of this might happen is considered in the next chapter.

12
Possibilities

This book is about history, albeit history with pressing relevance to the present and future. In this chapter I do not pretend to answer present problems, but I suggest what history might tell us about them. The first section is about the character and timing of the future transition away from fossil fuels; the second focuses on the first steps of that transition. The third section envisages ways in which longer-term and deeper-going changes to technological, social and economic systems might happen; and the final section considers who might begin to effect such changes. This is necessarily a concise presentation of huge, contentious questions; the reader's attention is drawn to the notes, which refer to books and articles that deal with them in more detail.

The character and timing of the transition away from fossil fuels

One way or another, society will move away from fossil-fuel-dominated systems. There are dystopian scenarios in which natural phenomena – the climate changes arising from global warming, resource constraints and so on – compel a collectively paralysed society to change. It is much more likely, though, that society will react, somehow, to the crisis it faces. This discussion of how it might do so starts from two questions. First, what is meant by 'transition' away from fossil fuels: a move to new technological systems, within existing social and economic ones – or the transformation, too, of those social and economic systems? Second, given what we know about global warming and other changes in natural systems, how long might, and/or should, the transition take?

Climate scientists have drawn parameters within which to discuss this second question: there is a consensus among them that humanity collectively has, in the period starting in 1870, a 'carbon budget' of about 3200 billion tonnes of carbon dioxide emissions to work with, if the most dangerous effects of global warming are to be avoided. At the current rate at which greenhouse gases are being emitted, this budget would be used up entirely, well before the middle of this century. (See pp. 57–8.) Beyond that, global warming effects (in particular, sea level rise, higher temperatures and weather volatility) could play havoc with humans' living conditions on an unprecedentedly destructive scale. These forecasts suggest the need for a transition away from fossil fuels within two or three decades – over a much shorter period than is usually envisaged. One damaging aspect of the public discourse around the UNFCCC process is that this contradiction is constantly ignored, downplayed or, worse, normalised.

What might the history of previous transitions from one energy system to another tell us about the character, and speed, of the transition now facing us? A recent research project[1] pointed to four conclusions.

1. That such transitions are 'predominantly characterised by changing types and amounts of energy end-use services' – that is, the way that energy is supplied depends on how it is used by final consumers. For example, industry's thirst for motive power (steam) drove the transition to coal in the early nineteenth century.
2. Technological innovations 'tend initially to be crude, imperfect and expensive'. Neither the steam engine in the nineteenth century nor the solar panel in the twenty-first could easily become cost-competitive with incumbent technology.
3. Technological change from innovation to widespread diffusion 'is generally slow, lasting as a rule many decades', and technologies come in clusters, with their infrastructure.
4. The 'transformative potential of energy technologies arises through clusters and spillovers', not through 'eureka' moments, implying, again, 'slower potential rates of change'.

In the foregoing chapters I have sought to build on conclusion (1). During the second Industrial Revolution, electricity for industry, and fuel for car-based transport systems, were among the drivers of fossil fuel demand; demand for these forms of useful energy was in turn determined by social and economic systems. In the late twentieth century, fossil fuel demand was driven by these, and other, technologies, underpinned by urbanisation, industrialisation, growth of mass consumption and so on. Much public discussion about future energy systems takes for granted that these trends will continue. It is often framed by ideas of 'progress', according to which ever-expanding US-style personal consumption (car-based cities, detached homes with multiple energy-guzzling possessions, and so on) and China-style industrial consumption (production of energy-intensive materials, such as steel and aluminium, with export of manufactured goods at great energy cost for energy-inefficient use), are inevitable, if not desirable. This, after all, is 'economic growth'. These ideologised assumptions need to be challenged. The prospect of changes not only to technological systems, but also to social and economic ones, opens up bolder and more attractive scenarios for the energy transition.

An energy transition could not leave untouched the inequality inherent in present social, economic and technological systems. Arnalf Grubler and Charlie Wilson, lead researchers on the programme referred to, argued that the post-second-world-war transition to an oil- and electricity-dominated system was 'incomplete', in that it left behind 1 billion people with no electricity access and a larger number with limited access.[2] But this transition was not directed at providing electricity access or improving lives; if we can speak of an aim or direction, it was to do with capital accumulation and the concentration of wealth and power; the inequalities were reproduced and deepened by the dominant social relations. A future transition that leaves these social relations intact, while switching technologies, will surely not tackle inequalities.

Conclusions (3) and (4) above continued a broad consensus among historians of technology, inspired by the work of Thomas Hughes and others, about 'technological lock-in' or 'path dependency' – the idea that technological regimes, once put in place and invested into, produce a 'congealed culture' of institutions and patterns of social organisation, that make them hard to shift.[3] Technological change, then, must be all-sided. '[T]echnologies, political and legal regulations, economies of scale and price signals, and social attitudes and values' must be changed together, the energy researcher Benjamin Sovacool argued recently.[4] As for timescales, Sovacool challenges the consensus, and argues that technological systems can change very quickly. He gives examples of systems that changed in one or two decades: energy-consuming technologies such as the energy-efficient lighting installed in Sweden's commercial buildings (1980s), or hundreds of millions of new cook stoves installed in China (1980s–1990s) and Indonesia (mid-2000s); and energy-producing technologies such as nuclear power in France, and CHP using coal and wind in Denmark (both 1970s–1980s). In wider public discussion, the rapidity of the technological transformation of communications, with the microprocessor, the mobile phone and the Internet, is cited.

To what extent can we generalise from Sovacool's examples? Certainly they show that technological systems can change quickly – if governments act decisively, as they did in all these cases. Obviously, though, a future transition away from fossil fuel-based technologies on a world scale will be not only geographically more widespread but qualitatively much deeper. France adopted nuclear electricity generation without tampering with systems of consumption (technological or social/economic) more generally; China changed its cook stoves without even starting to shift coal from its position as the main fuel; Denmark accomplished a more systemic change in both primary and secondary energy, but only at relatively small scale. The microchip-based communications revolution is a better example of a global shift, but a move away from fossil fuels would inevitably be much deeper-going. There are far greater quantities of physical technologies used for their production (oil fields), transformation (power stations) and consumption (city infrastructures, industrial and agricultural processes); and fossil fuels are much more deeply embedded in the economy than pre-microchip communications technologies.

Conclusion (2) above, that markets by themselves have not, and may not, rapidly take new technologies from innovation to diffusion, is significant with regard to both the timing, and character, of a future transition. This was emphasised at a gathering to discuss the research in 2011 in Cardiff, Wales. In past transitions, there had been 'clear private benefits' for producers and consumers in switching to new technologies, whereas such benefits are less obvious in the case of low carbon energy sources and technologies. For this reason, some attendees argued, a low carbon transition 'will need to be deliberately managed or engineered'; governments would have to act; the role of private business interests that control the fossil-fuel-based system would be 'critical' because 'the losing incumbents are likely to "fight back"'.[5] The obstructions to the future transition are political and social, more than technological, in other words. Related conclusions have been reached from research of technological scenarios for the shift away from fossil fuels. The latest of several such studies found it 'technically and economically feasible' for energy globally to be supplied entirely from renewables by 2050, while 'the main barriers are social and political'.[6]

How will such barriers be tackled? How will resistance of 'losing incumbents' be overcome? Will governments move against them? Where governments have sought to 'manage or engineer' energy transition, as in Sovacool's examples, the motivation has rarely been to do with global warming, and more often about, for example, fuel supply options or air quality. Faced with global warming, governments have failed, collectively: they assigned the main role to markets, despite the compelling evidence that in the absence of 'clear private benefits' that would not work, while continuing to subsidise, and politically support, incumbent technologies.

Those who offer strategies for the transition away from fossil fuels generally have two types of answer to this dilemma. The first was proposed by the Stern Review and embraced by international organisations (including the IEA, OECD and World Bank) and much mainstream commentary: markets *will* do it, provided they are regulated correctly; technological systems can be changed by redirecting, but not altering, social and economic systems. The continuing absence not only of regulation, but even, in most governments' cases, of any serious intention to regulate – and even the election of a pronouncedly anti-regulatory president, Donald Trump, in the USA in 2016 – has done little to dispel such credibility-straining optimism. It becomes, then, a justification for the status quo. A second type of answer is that a forced march is needed – 'planned economic recession', in the view of Kevin Anderson and Alice Bows of the Tyndall Centre, the UK's foremost climate change research hub. Some writers have used the analogy of the mobilisation to defeat fascism during the Second World War; socialists and radicals have engaged with that critically,

First steps of the transition away from fossil fuels

The strongest evidence that the transition away from fossil fuels is underway is in the electricity sector, where renewable technology has in the last decade become the largest recipient of investment. Renewables' share of world electricity generation has risen to about 5 per cent.[8] Renewables' costs continue to fall – to plummet, in the case of solar panels – putting them in a stronger position to compete with fossil fuels in some contexts. In Denmark, Spain and Germany, wind power supplies a substantial share of electricity.

Advocates of the UNFCCC process routinely claim that, as technology continues to improve and renewable costs continue to fall, market forces – hopefully helped by a reversal in the fortunes of failed carbon pricing – will push out gas and coal, and pull in wind and solar. Two sets of factors raise doubts about whether, and how quickly, this might happen. First, electricity markets as currently designed cannot cater easily for renewables, which have a high up-front investment cost (for building wind turbines, installing solar panels, and so on) but a negligible operating cost (i.e. once the equipment is in place the electricity is produced almost for free). Experience shows that once there is a substantial proportion of renewables in a particular market, healthy gusts of wind and sunny days can drive down wholesale electricity prices and even make them collapse. 'The more renewables there are in the system, the more often such collapses occur', The *Economist* observed. Specialists in the field propose changing market rules to favour smaller and renewable generators, which fossil fuel incumbents will not welcome.[9]

The second set of problems is technological. Renewables are intermittent, i.e. not constantly available at the same level because they depend, for example, on the sun shining or the wind blowing. One set of answers to this lies in the development of energy storage such as in batteries or hydrological systems. Batteries are much more efficient than they used to be, but not yet capable of dramatic transformations; and there are questions about the availability of materials that would be needed for much more widespread manufacture. A second set of solutions to intermittency lies in integrating renewable generation into electricity networks that are: decentralised (in industry terminology, distributed generation); technically capable of accepting electrical power from a multitude of smaller generators; and flexible, meaning that the grid can manage demand to reduce peaks (when demand goes up, pushing the system's capacity to its limits), adapt some demand to supply (diminishing non-urgent demand when intermittent supply is lower), and eliminate waste. Decentralised grids were already made desirable from the engineering point of view in the 1980s, by the reversal of the trend to ever-bigger power stations, and made possible by the revolution in computer technology around the same time.

These issues are at the centre of discussion among electrical engineers. Some see great scope for 'prosumers', who produce electricity for their own consumption and feed any spare back into the grid. A recent controversy among researchers, about the lessons of large-scale renewables deployment in Spain and Germany, produced, first, warnings that, without stronger government support, renewables' share of electricity generation will never move from a significant minority to a majority; and, second, doubts about overcoming technological barriers to 100 per cent renewables generation.[10]

With regard to transport, much public discussion about the transition away from fossil fuels has focused on electric cars. Of a global vehicle population that passed 1 billion in 2010 (including cars, trucks and buses), close to 1 million (around 0.1 per cent) are electric. In rich country cities in particular, the numbers are fast increasing; electric cars may within a few years become as cheap as ICE-powered cars.[11] In the view of particularly extreme 'techno fixers', electric car prices will fall so steeply during the 2020s that they will become ubiquitous, dealing a death blow to the oil industry and forcing the motor industry to reinvent itself. Such claims (i) ignore the fact that cars account for a chunk of global oil demand, but still less than half; (ii) ignore the inertia of 'infrastructure lock-in' (that is, assume that the array of petrol stations, vehicle repair businesses and other support for cars could be completely replaced in a few years), and (iii) discount the opposition of car manufacturers and governments that have consistently refused to use regulation to reduce car use. Nevertheless, much wider diffusion of electric cars seems likely.

But transition away from fossil fuels implies a much more profound transformation of transport. Electric cars fit neatly into the 'green growth' narrative, but their potential for cutting fossil fuel use is limited. First, they are only as 'green' as the electricity they use: although electric motors are more efficient than ICEs, running cars on fossil-fuel-generated electricity may not reduce overall consumption. Second, 'electric cars suffer from the inherent inefficiency of all personal motorised, road-based transport: the need to move a one- to two-tonne vehicle in order to transport a few hundred pounds worth of people', as the energy researchers Richard Heinberg and David Fridley put it. Third, electric cars do nothing to reduce the quantities of fossil fuels consumed in making roads, parking spaces and the cars themselves. An effective transport transition will involve great changes to cities, where most of the world's car journeys are undertaken. Other modes of transport – public transport, bicycles, walking, trains, and so on – will become dominant, and urban infrastructure be geared to them.[12]

In contrast to the steps taken in the electricity sector, and the tiptoe steps in transport, the transition away from fossil fuels is hardly at the beginning of the beginning with regard to industry, and urban infrastructure more generally, including construction and heating of buildings. For rich countries, the export of energy-intensive industrial processes to developing countries has largely put them 'out of sight, out of mind' in policy terms. Neither the technological difficulties of producing non-fossil high-temperature heat for industrial processes, in the first place to make steel, aluminium and cement, nor ways of reducing demand for such materials, have been faced. Little has been said about either substituting for fossil fuels in the production of petrochemicals, fertilisers, asphalt and other by-products, or rethinking their uses.[13] Much is made in public discussion about reducing households' final consumption. But much of this consumption is either indirect, or non-discretionary, or both. Households are unlikely to make drastic changes to their consumption habits that would make a noticeable difference, without equally drastic changes in the economic, social and cultural contexts in which they live.

To sum up. The growth of renewables in electricity generation is a real, and significant, indication of the potential of non-fossil energy supply technologies. They have so far taken root only in small corners of the electricity system; claims that they have pronounced a death sentence on coal are as yet exaggerated. In the other major sectors – transport, industry, urban infrastructure, buildings and households – only negligible progress away from fossil fuels has been made. Transition means changing not only the way that energy is produced, but also the technological systems that consume it and the social and economic contexts in which they operate.

'Green growth' advocates claim that market levers can be used to push the transition, notably, the idea of 'stranded assets'. This idea states that as the transition away from fossil fuels gains momentum, assets owned by fossil fuel producers (oil and gas resources, and field licences) or electricity generators (fossil-fuel-fired generating equipment) will lose value. If oil, gas and coal must be kept in the ground in order to prevent dangerous global warming, the argument goes, then values attributed to companies on the basis of their reserves of these fuels must be reduced. The idea of 'keeping it in the ground', proposed by environmentalists, has received mainstream attention. In 2015, for example, the Governor of the Bank of England said that the 'vast majority' of fossil fuel resources would be 'stranded' if policies were adopted to limit global warming to 2°C.[14]

Clearly, if demand for fossil fuels falls over the long term, the value of assets based on them may fall; equally clearly, there is as yet no sign of such a long-term trend. Values attributed to fossil fuel companies by the market are influenced firstly by fuel prices; prices in turn are influenced by market perceptions of supply and demand; these influences are not direct, and it is rarely possible to distinguish the parts played by shorter- and longer-term factors. In 2015–16, for example, due to falling coal demand in China and the USA, market prices fell sharply, and coal companies' share prices followed, resulting in some bankruptcies. (See pp. 167–8.) Some optimistic commentators rushed to proclaim the beginning of the end of coal, but this is as yet only possible, not certain. Scientists' preliminary figures for 2017 suggested that coal consumption growth had resumed again.[15] In the case of oil, prices have in the last decade stayed at a much *higher* level than previously. This has not only increased the perceived value of resources in the ground, but also made it economically feasible to recover oil that is technologically more challenging to produce, thus opening up new sources of supply. The conclusion is not that assets cannot be stranded, but that it is important not to confuse short-term market phenomena with long-term trends, and not to confuse hope with reality.

Part of the stranded assets argument relies on the prospect of regulation, brought about by policy changes. If governments fulfil their promises to reduce greenhouse gas emissions, the argument goes, fossil fuel assets will more rapidly be stranded. Stock markets are sceptical: the announcement of emissions reduction targets at the 2015 Paris climate talks had no discernible effect on fossil fuel companies' market values, because it did not change markets' view that regulation is unlikely to threaten asset values for the foreseeable future. Governments *have* taken actions that may help the movement away from fossil fuels, but these have been limited (such as, support for renewables in Europe), sometimes responsive to fears of social protest (like controls on coal-fired power in China, due to urban pollution issues) – but mostly overshadowed by political support for fossil fuels.[16]

It is politically important to the main participating governments that the UNFCCC process and its market mechanisms, supposedly the single means to address global warming, be seen to be working. This creates great pressure, through mainstream commentary, to paint it in brighter colours than it deserves; and unduly distorts discussions of stranded assets.

This pressure should be resisted. We should acknowledge the failure of the quarter-century-long process; soberly assess timescales; and base our hopes for the future on realities, not illusions.

Prospects for social, economic and technological change

Potential future changes to technological, social and economic systems, in the course of a transition away from fossil fuels, may be grouped in three types.

First, there are *changes to, or adaptations of, existing technological systems that could reduce fossil fuel use rapidly*. One such change, deployment of electricity generation from non-fossil sources, has begun. Other technologies have for decades amounted to unrealised potential – what Amory Lovins in the 1970s called 'roads not taken': ways of changing energy transformation and distribution (such as, CHP and decentralised electricity grids), or of conserving energy in consumption (for example, in industrial processes, construction methods, or car manufacture and fuel use). Abundant reports from official bodies and researchers contain details.[17] At numerous turning points, such changes could have been made, and were even under discussion by political and business elites, but either did not happen or happened only in a watered-down form. Experience shows that governments can either push, or obstruct, such changes; civil society can also intervene.

A second type of change, which borders on and overlaps with the first type, would amount to *superseding the technological systems in their current form*. These systems are only likely to be dislodged as a result of far-reaching social change and, where necessary, breaking the resistance of incumbent interests that control them. Four such changes are:

1. Remaking the relationship between cities and countryside, by making the divisions between them less extreme, and moving urban built infrastructure away from the currently dominant energy-intensive model. This could end construction of energy-inefficient buildings using energy-intensive materials, and cut sharply the demand for fossil-fuelled space heating and other energy-intensive practices. Land ownership patterns would have to change.
2. Transforming urban transport infrastructure, (a) to decentralise goods distribution, and (b) to gear cities and towns to public transport, bicycles, walking and other modern transport technologies, thereby superseding the age of car-based transport systems. This would involve overcoming the resistance of corporate interests (fossil fuel, car, steel and aluminium makers, road builders, and so on).
3. Moving to fully integrated, decentralised electricity networks, geared to multiple small electricity producers, managed by 'smart' technology, thereby reducing or ending the need for centralised fossil-fuel-fired power stations. This would be (is being) resisted by electricity companies.
4. Changing the character of widely diffused energy consumption technologies (household heating and electricity systems), and other products, to make them repairable by users, thereby reducing waste, overproduction and the effects of planned obsolescence – changes that are incompatible with current profit-based marketing.

These deeper-going shifts, involving technological change together with social and economic change, point towards the third and most thorough-going type of change, the *transformation of the social and economic systems that underpin the technological ones*. We can envisage forms of social organisation that supersede corporate and state control of the economy, advance collective and community control, and, crucially, in which employed labour – a central plank of profit-centred capitalism – is superseded by more meaningful types of human activity. This is what I understand by socialism: a future social form antithetical to twentieth-century state 'socialisms'.[18]

Changes associated with this could include:

1. Transformation of productive activity (industry) beyond the constraints of the wage labour system, and corporate control, unleashing human creative capacities to make things that are truly useful and desirable. Such production, coordinated with twenty-first century information technology, would supersede the production of little-needed or unneeded goods, waste in processes, and the egregious use of energy-intensive materials.
2. The continued transformation of domestic labour with energy-efficient technologies.
3. A transition away from industrial agriculture, reversing the tendency towards the production and transportation of fossil-fuel-intensive meat and luxury foods for a minority at others' expense; and reversal of the trend towards long-distance and international bulk transportation. This could enhance human health, minimise waste and slash fossil fuel inputs.
4. A society where employed labour is superseded by useful and creative activity (production for use) could move away from consumerism and ideas that material goods are essential means to happiness and fulfilment.

These are speculations, not blueprints. Their purpose is to indicate the gigantic potential opened up by thinking about the transition away from fossil fuels in the larger contexts of developing sustainable technological systems better to meet human need, and changing technological systems together with social and economic ones.

Several writers concerned with radical social change (Paul Mason, Nick Srnicek and Alex Williams[19]) have argued that technological changes begun by the third Industrial Revolution will smooth the way to a post-capitalist society, by democratising work processes, reducing the working day, and so on. The argument needs to be put the other way round. History shows that technologies, as they develop in definite social contexts, are shaped and constrained by those contexts. The development and diffusion of technologies that will enhance the transition away from fossil fuels have been, and are, constrained by dominant social and economic interests that stand to lose out. While there is no simple set of rules governing the relationship between social, economic and technological transformations, the logic of automation-to-post-capitalism must be turned on its head: the full potentials of technology will only be unleashed, freed from these constraints, by social and economic transformations that move beyond capitalism.

Forces for change

The historical fact that the UNFCCC process has failed to produce reductions in fossil fuel consumption suggests that more plausible agents of change are social forces *outside* the climate talks, and outside the participating governments. This case was made by Philip Bedall and Achim Brunnengraber:[20]

> When the entire carbon cycle is considered, from the extraction of fossil fuels to the harmful emissions, the area of conflict of climate politics is extended. From this wider perspective, a transformation of the organisation of the energy systems and the way of life and the mode of production becomes necessary. This perspective necessitates changes that go far beyond a technological modernisation or economic measures such as emissions trading. The debates on a 'great transformation', a Green Economy, or a post-growth society and Prosperity Without Growth refer to these requirements.[21] But *who are the actors to initiate such a transformation towards a new social system*, [my emphasis, SP] if the UN's negotiations seem too narrow thematically to achieve it, and beyond that confronted with a process of renationalisation [i.e. the return of decision making on climate issues to national governments] avoiding any substantial adjustment?

The actors referred to can only be civil society in the widest sense. History provides not only endless examples of social action producing change in the broadest political sense, but also evidence that such movements can shape systems that produce and consume energy. In the 1970s and 1980s, social opposition was crucial in blocking the progress of nuclear power, political elites' favoured successor to fossil fuels. Social movements have shaped electrification processes, as in South Africa and India. And, at least in Europe and China, public concern with air quality has hastened the decline of coal. Even where civil society seems passive or has no organised expression, governments make policy decisions out of fear of its anger, by, for example, providing electricity as a state benefit. Obviously, the transition away from fossil fuels is a far deeper-going process than any of these; it must be made the concern of society as a whole, in order that society as a whole can participate in making it happen.

An obvious objection is that this sounds like a long process, and that the timescales set by climate science do not allow that luxury. An obvious riposte is that we have 25 years of evidence that this problem is beyond the capacity of political processes to resolve. Together with that, we have very clear evidence of how the world's most powerful governments are likely to react to the effects of global warming as it advances. Here are three examples:

1. The flooding of New Orleans in 2005 during Hurricane Katrina. It is not known whether this was caused by global warming, but it is known that such extraordinary weather events are becoming more common as a result of global warming. Although the victims were citizens of the world's most powerful country, hundreds perished and thousands lost everything, sharing the fate of the much more numerous victims of natural disasters in countries outside the rich world. Hopes were expressed at the time – and again during the extraordinary storms in the USA in 2012 and 2015, and the flooding of southern states in 2017 – that these events might push the political elite into revising its stance on global warming. They did not.
2. The effects of hotter weather and flooding on tropical-zone agriculture in recent years. Global warming is among the underlying causes of the hardships, in many cases fatal, visited on agricultural communities in Africa and southeast Asia as a result.[22] The international political reaction has been one of indifference.
3. The sharp increase in 2016–17 in the number of refugees travelling to Europe from the Middle East and North Africa. This was caused primarily by military conflict, not global warming – but it prefigured the type of migratory movements researchers see as likely to be caused by future climate change. The grim response of many European governments – that in various degrees allowed refugees to drown en masse, confined them to detention camps, and used them as a cynical bargaining chip in geopolitical haggling with the government of Turkey – indicates how they might deal with future effects of global warming: by building walls, real (as commended by president Trump) or virtual.

So we do not need to imagine the future international political response to global warming effects. The prototypes are here: handing the transition away from fossil fuels to the markets, with disastrously ineffective results; and cordoning off the rich world from the most violent and damaging consequences. For civil society to take matters into its hands cannot possibly be a simple or easy answer – but it can and will find better answers to problems than these. In a study of social responses to disasters, the writer Rebecca Solnit demonstrated convincingly that people tap into resources of solidarity and collective action that they might not have known they had.[23] The global warming disaster – which has been produced socially, on a much greater scale and over a much greater time span, than other disasters – demands a collective response from us all that we, too, might not have thought possible.

13
Conclusions

Fossil fuel consumption has since 1950 expanded much more rapidly, and in a qualitatively different way, from its growth previously. The Industrial Revolution of the early nineteenth century provided a crucial stimulus to coal consumption, giving it a central place in economic activity for the first time. The second Industrial Revolution of the late nineteenth century (and specifically, electricity generation, the internal combustion engine and artificial fertiliser production) provided the technological basis for a proliferation of economic activities to which fossil fuel consumption was integral. After 1950 these activities expanded, becoming dominant in the rich countries and, from the 1980s, elsewhere. The rapidity with which fossil fuel consumption has grown fits well with the idea of a 'great acceleration' of economic activity from the mid twentieth century.

People consume fossil fuels through technological systems that are embedded in social and economic systems. Interpretive frameworks that isolate consumption from these systems, and/or isolate consumption from production, are misleading. Because fossil fuel consumption growth is driven by these systems, it often correlates with economic trends, but not with population growth. In spite of this, the assertion that population growth is a key cause of fossil fuel consumption growth is repeatedly endlessly, reflecting populationist ideologies that see the number of humans, rather than the social and economic relationships in which they live, as the cause of the dislocation between humanity and nature. The historical record shows that the relationship between population growth and fuel consumption levels is indirect and mediated through systems, and that changes in the systems are the main determinants of consumption levels.

Much sociological research on consumption – not only of fossil fuels but of goods in general, mostly in the rich countries – stops short of explaining the relationships between individual consumption and these systems. Social habits and routines, and the influence of consumerism, need to be considered in the context of these systems. An example is the distinction between discretionary and non-discretionary consumption of fossil fuels, which much research ignores. Part of the problem is the division of disciplines in academia: some historians and researchers of society, who focus on individual consumption, pay insufficient attention to the work of engineers and natural scientists, who focus on inefficiencies in technological systems, and vice versa. Analysis needs to be unified.

Since the Industrial Revolution, energy carriers have increasingly been supplied as commodities, via markets. Some of them, particularly electricity, have alternatively been treated as a state benefit, to spur economic development by supporting industry and urban populations. Such provision came to be seen as a right, one of the welfare benefits associated with the rise of the labour movement and of forms of state socialism. Views of electricity as a commodity, or as a right, clashed repeatedly: the liberalisation of the 1990s pushed back, but did not destroy, the idea that electricity is a right. The commercial energy system, in which this opposition persists, has never reached large sections of the world's population, particularly rural communities outside the rich world, which rely on non-commercial energy systems based primarily on biofuels.

The chief driver of fossil fuel consumption growth since 1950 has been economic expansion, and in particular the expansion of production, driven by the constant impetus to capital accumulation. In quantitative terms, the only interruptions in the constant upward trend of consumption have been major economic crises (in the mid and late 1970s, and in 2008–9). The most significant trends through which economic expansion has driven fossil fuel consumption growth have been industrialisation, the transformation of the labour process, electrification, urbanisation, motorisation and the growth of mass material consumption and consumerism.

In the rich countries, broadly speaking, during fossil-fuel-intensive post-war industrialisation, consumer goods that were previously luxuries became affordable for millions of working people. From the 1980s, people tended to work longer hours to pay for fossil-fuel-intensive cars, houses and appliances. Cheap fossil fuels and electricity changed the labour process both in industry and in the home, where electric refrigerators, washing machines and vacuum cleaners, as well as gas cookers, lightened the physical load of domestic labour. But electrification, like other technologies, has made little difference to the hours worked, whether by employed or domestic labour.

Motorisation provides a graphic example of how technological, social and economic systems shape fuel consumption trends. In the USA from the 1920s, government and car manufacturers deliberately undermined railways and public transport to ensure the dominance of private cars and trucks over the whole transport system; infrastructure spending favoured roads; this state support was complemented by marketing techniques (heavier cars, planned obsolescence). Car-based urban transport and fossil-fuel-intensive buildings combined in fossil-fuel-intensive urbanisation, which was generalised across the rich world in the post-war boom, and parts of other countries from the 1980s – despite the dangers of global warming having been confirmed by then.

Electricity networks, which accounted for one tenth of fossil fuel consumption in 1950 and one third by the 2010s, were shaped by social and economic changes. Electricity infrastructure was mainly constructed by the state. Private companies nowhere implemented rural electrification; privatisation in the 1990s did not advance it. In the Soviet Union and China, as in capitalist countries, electrification of towns and industry took priority over the countryside. The way that social and political forces shape technological systems is graphically illustrated, as in varying patterns of electrification in different Indian states.

In the post-war boom, when the rich countries' economies expanded more rapidly than at any other time, fossil fuel inputs, cheap relative to the cost of labour, played a fundamental role. The 1970s oil price shocks rebalanced the oil market away from the state-supported rich-world oil companies towards producer nations, and put an end to the assured supply of cheap energy. They coincided with economic crises that form a turning point in the history of fossil fuel consumption. Thereafter, consumption increased more slowly in the rich countries, due partly to efficiency improvements in industry but, more significantly, to the export of energy-intensive industries to developing countries. From the 1980s, fossil fuel consumption grew more rapidly in the latter, culminating in the large increases during the Chinese industrial boom of 2003–8. At every stage, technological opportunities for energy conservation were limited almost entirely to those that coincided with the private interests of industrial fossil fuel consumers; potential for electricity network decentralisation, and conservation in industrial processes, was obvious technologically, but left unrealised.

Increased fossil fuel consumption outside the rich world from the 1980s was accompanied by the appearance of forms of mass individual consumption that were already ubiquitous in rich countries. But it did not reduce the inequalities inherent in systems of consumption, which reflect those in society as a whole. More than 1 billion people remained without electricity access, and an even larger, and growing, number had limited and/or sporadic access. Most people in both groups relied, and rely until today, on non-commercial biofuels for cooking and other essentials. Those living in the large grey area between commercial and non-commercial energy systems – very often, the urban poor outside the rich world – have often framed access to electricity as a human right, in opposition to attempts to introduce it as a paid-for service, and participated in social movements demanding access.

Some further conclusions concern the transition away from fossil fuels. The discovery of global warming in the 1980s provided an imperative for this transition quite unlike anything previously. The greenhouse effect, by which fossil fuel consumption and other economic activities are directly causing a rise in global average temperature – which in turn causes sea level rise, volatile weather and other physical changes inimical to society – is the most extreme result of a dislocation, or rupture, between human society as currently organised and the natural environment. The scientific consensus is that action should be taken to avert an average temperature rise above 1.5°–2°C higher than pre-industrial levels (although the perception of 'safe' limits, which involves assumptions that are not universally shared, can not be normative). The global carbon budget available to meet these targets will be used up in one or two decades at current rates of consumption.

The international talks since Rio have completely failed to make any progress in terms of reducing fossil fuel consumption. The widening gap between the targets referred to in agreements, and the practical outcomes of 25 years of international diplomacy, speaks volumes. Underlying this is a major crisis for the leading states: they ground their claims to legitimacy in terms of representing society as a whole, but have clearly abandoned the common social interest to the private interest of elites. International action under the Kyoto protocol (1997) was limited to market schemes that have been a discredited failure; in the 2000s, some timid elements of regulation have been added, but their effect has been dwarfed by the overall increase in fossil fuel consumption. Meanwhile, a huge industry has sprung up, producing reports and advocating solutions for the talks, as they have drifted further and further from a solution. The Copenhagen (2009) and Paris (2015) conferences were the culmination of this process of failure. An important function of analysis must be to separate the reality of policy outcomes from the discourse around the talks that acts as a lever to constrain discussion about global warming within the framework of market-based solutions.

The limited progress made in reducing fossil fuel consumption has been due mainly to (i) some state regulation, e.g. in Europe, Japan and China, and (ii) the development of renewables for electricity generation. On the other hand, negligible progress has been made in reducing emissions by transport, industry, construction and agriculture. The true measure of governments' collective commitment to reducing fossil fuel consumption is not the declarations made during the Rio process, but the level of subsidies to both production and consumption, that run into hundreds of billions of dollars per year, and rose in the 2000s.

The transition away from fossil fuels is not solely a technological issue. Energy-intensive technologies have been privileged over less energy-intensive ones (such as, cars over buses and bicycles), not because the latter do not exist, and sometimes not because the latter are less economically viable, but for a complex of social, political and economic reasons (in this case, the centrality and lobbying power of the car industry, trends in urban development, and so on). Similarly, the production of commercial energy from fossil and nuclear fuels has been privileged over renewable technologies, and not for technological reasons, as the example of solar power's 'false start' in the 1970s shows. Fossil fuels have had not only the advantage of incumbency, but also the political support implicit in the high level of subsidies.

The third Industrial Revolution in computer and communications technologies, since the 1980s, showed how social and economic systems constrain technological potential. On one hand, the possibilities for Internet-type technologies to conserve energy by transforming management of, and enhancing decentralisation of, electricity networks, have not been tapped. On the other hand, the expansion of the Internet under private commercial control has created a major new source of avoidable, wasteful electricity consumption.

There are no simple or obvious formulas for hastening the transition away from fossil fuels. The history of attempts to do so indicate that, first of all, progress lies *outside* the framework laid down at Rio, which assumes that solutions must be found within the narrow confines of 'economic growth' dogma. Transition requires questioning the assumptions that political and social elites are able to solve these problems at all, since clearly so many of them have a vested interest in constraining solutions within frameworks on which their wealth and power rests. The reaction of these elites to social and economic disasters that are either related to global warming, or are similar to the problems it will undoubtedly cause (flooding, agricultural problems and flows of refugees) indicate how they will react to future crises caused by warming: by reinforcing their own control.

Radical social change that will produce the most favourable contexts for a transition implies that society will take its future into its own hands and out of the hands of political and economic elites. Visions and programmes need to be developed for the transformation of technological systems together with social and economic systems; not separately from them.

Appendices

Appendix 1
Measuring environmental impacts, energy flows and inequalities

Researchers of the history of fossil fuel consumption face an array of statistics and other numbers, measuring not only levels of consumption, but also the impact of fossil fuel use on the environment (primarily, but not only, via CO_2 emissions), the way that fossil fuels are used in technological processes, and inequalities in forms of consumption. All of these numbers highlight some realities, and push into the background, or omit, others. Those most relevant to this book are outlined here.

Measuring environmental impacts

With the growth of environmentalism in the rich countries from the 1960s, academic researchers began systematically to try to measure the negative impact of human activity on the natural environment. In 1970, a controversy erupted on interpretive issues between the biologist Paul Ehrlich (a neo-Malthusian advocate of population control) and ecologist John Holdren on one side, and the plant physiologist Barry Commoner on the other. Commoner sketched some calculations covering the US economy in the post-war period, from which he concluded that the main cause of atmospheric pollution was technological change, rather than rising population or rising living standards. His examples included pollution from electricity generation, nitrogen fertiliser manufacture and chemical wastes. Perhaps the most memorable concerned the introduction of the non-returnable beer bottle, which meant that in the USA between 1950 and 1967, population rose by 30 per cent, beer consumption per capita rose by 5 per cent and the number of bottles used per unit of beer shipped rose by 408 per cent. It was the new bottles that strained resources far more than the increase in population or in the amount of beer being consumed, Commoner wrote. This, he argued, supported his view that the misuse of technology, not population growth or rising affluence, was the primary cause of damage to the environment.[1]

Ehrlich had appealed to Commoner privately not to publish his results, on the grounds that they would obstruct the campaign for population control. Commoner rejected this indignantly, and a bad-tempered polemic followed.

Ehrlich and Holdren argued that the impact of population growth on the environment was 'disproportionate', and outweighed other factors; Commoner insisted that 'the technological factor' was far more significant, and that Ehrlich's view had been skewed by his fervent populationism.[2] Ehrlich took an initial formula with which Commoner had made his estimates, and rewrote it as:

Impact = Population x Affluence x Technology

This formula, known as IPAT, served as the basis for many subsequent attempts to measure numerically damage to the natural environment by human activity.[3] Researchers developed versions of it using scales such as GDP per head to measure affluence, and scales such as the intensity of greenhouse gas emissions in economies to measure technology. These 'top down' statistics can only reflect complex processes in limited ways, and miss many economic and technological phenomena uncovered by 'bottom up' investigations. Such methodological problems have long been obvious. Applying the equation 'has led to the elementary logical error of aggregation, whereby calculations are made at the global level, without concern for whose population is increasing, whose consumption is growing and who is benefiting from technological change', a development researcher pointed out in 1992.[4]

Notwithstanding such warnings, much research made an *a priori* assumption that consumption by individuals – as opposed to social, economic and technological systems – constituted the best framework for studying environmental impacts.

In the early 1990s, researchers working with the IPCC sought ways of measuring the drivers of greenhouse gas emissions, rather than environmental impacts more generally, the subject of the Ehrlich-Holdren-Commoner controversy. From their discussions emerged the following equation,[5] based on IPAT and named the Kaya identity after Yoichi Kaya, the energy researcher who proposed it:

Carbon dioxide emissions = carbon dioxide emissions per unit energy x energy per unit output x output per capita x population.

A great deal of research on drivers of greenhouse gas emissions, including publications relied on by the IPCC, has used the Kaya identity as a starting-point, to support, for example, the headline conclusion in the most recent IPCC report that 'economic and population growth continue to be the most important drivers of increases in CO_2 emissions'.[6]

During the 1990s and 2000s, researchers in the field of human ecology developed a critique of the IPAT formula and its derivatives, including the Kaya identity. Thomas Dietz and Eugene Rosa argued in 1994 that, despite its 'appealing features', IPAT had 'serious limitations', and above all lacked 'an adequate framework for disentangling the various driving forces of anthropogenic environmental change'. They warned against drawing 'strong conclusions' on the relative importance of population, affluence and technology 'despite the paucity of strong evidence', and took aim in particular at a joint study by the US National Academy of Sciences and the Royal Society of London that asserted that population growth was 'a major threat to human well-being'. Donella Meadows, one of the authors of the influential *Limits to Growth* report (1972), in 1995 pointed in alarm to the IPAT formula's capacity for encouraging crude populationism. She quoted approvingly campaigners who denounced IPAT as a 'bloodless, misleading, cop-out explanation for the world's ills' that 'points the finger of blame at all the wrong places' and 'leads one to hold poor women responsible for population growth without asking who is putting what pressures on those women'.[7]

Dietz and Rosa proposed a reformulation of IPAT as a stochastic model (a mathematical method for producing randomly determined patterns that may be analysed statistically but may not be predicted precisely). Such a model, rather than claiming to show the relative importance of the driving forces, could underpin research on each of the factors that could better illustrate their significance. Dietz, Rosa and their colleagues used this method, given the acronym Stirpat (Stochastic impacts by regression on population, affluence and technology) in studies of the drivers of greenhouse gas emissions, employing increasingly complex computer modelling techniques.[8]

In a later article (2012) surveying two decades of work, Dietz and Rosa again warned of the limitations of IPAT and the Kaya identity. These equations cannot be used to test hypotheses about the relative contributions of drivers of emissions, they argued. The built-in assumption that the elasticity of emissions to each factor is the same is 'questionable'; and 'they do not take explicit account of culture and institutions'. This lack of attention to social forces by those trying to quantify the physical impacts of human activity on the environment seems to be exacerbated by the divisions between academic disciplines. Studies of those physical impacts often 'pay little or no attention to social systems and social networks', the environmental sociologists Artur Mol and Gert Spaargaren concluded.[9]

Measuring energy flows

In the initial controversy between Commoner, Ehrlich and Holdren, the two sides had disagreed not only about the interpretive framework, but also about a significant detail: how much more energy would be used to make something out of aluminium than to make it out of steel. Commoner believed aluminium production was 15 times more energy-intensive than steel production; Ehrlich calculated it was five times. Energy researchers took up the challenge: Bruce Hannon returned to Commoner's example of non-returnable beer bottles, and found that they used 3.11 times the amount of energy per unit of beer as returnable bottles. It soon became clear that there were no agreed criteria for such calculations, and no academic tradition to fall back on.[10]

In the USA, with the 1973 oil shock focusing political attention and research funding on conservation, it was deemed urgent to formulate conventions for measuring energy flows through economic systems. One method developed as a result was net energy analysis, whereby researchers draw the boundaries of a system (such as, a manufacturing process or branch of industry), and measure energy flows through that system. Howard Odum pointed to the method's potential with an article showing that stripper oil well operations used more energy than they produced. This was part of the groundwork that he laid for holistic analyses of the flow of materials and energy through both natural and social systems.[11]

Methods similar to input-output analysis, which was first developed by the economist Wasily Leontief in the late 1940s, were adapted for net energy analysis. Input-output analysis studied the relationship of different systems and sectors in the economy by measuring flows into and out of them by their monetary value. Some early studies applied this method to energy systems.[12] Then a team at the University of Illinois began to measure, instead, in energy units (i.e. joules) rather than money – drawing criticism from economists who argued that energy units might no more accurately reflect economic processes than monetary ones.[13]

Research funding for net energy analysis and associated methods in the USA suffered as a result of president Reagan's attack on environmentalism. After his Executive Order (1980) that required regulatory action to be justified in economic, rather than net energy or environmental, terms, funds went elsewhere. But internationally, such research methods were boosted by the Brundtland Commission (1987) and the acceptance of sustainability as a political principle. Daniel Spreng published an important overview of the application of net energy analysis in 1988. Thereafter, net energy analysis methods were included in larger holistic studies of flows such as life-cycle analysis to track the origin of greenhouse gas emissions; and material flow analysis, applied both to energy carriers and to other materials used in the economy.[14]

The new discipline of environmental sociology often combined qualitative research on social change with quantitative methods that bridged economics, engineering and the natural sciences. Flow analysis became increasingly sophisticated in sub disciplines including industrial ecology and environmental systems research.[15] More recently, net energy analysis has been combined with process analysis in industry, which asks which goods and services are required to produce a product, distinguishes between energy and non-energy inputs, and examines the non-energy inputs in turn to see what energy inputs it required.[16]

Since the 1990s, researchers have applied net energy analysis, or similar techniques, to household energy consumption, to investigate how differences in income, between rural and urban households, and other differences, impact consumption. An important distinction was drawn between *discretionary* and *non-discretionary* consumption, i.e. between energy that individuals decide to consume, and energy consumed in the course of flows outside the consumers' control.[17] Unfortunately this distinction has yet to gain wide acceptance, possibly, in part, because of the division between academic disciplines – in this case, insufficient communication between the engineers and economists who dominate quantitative research projects on consumption, and the environmental sociologists who developed the idea.

Measuring inequalities

The adoption of the UN Framework Convention on Climate Change (UNFCCC) in 1992 was accompanied by diplomatic and political disputes over the principles that would guide future emissions reductions, and possible financial compensation for past emissions, to be paid to developing countries by rich countries. Against this political background, the calculation of past, present and future emissions per head of population assumed great importance. A consensus eventually formed around estimates calculated by academic researchers from international organisations' statistics. The figures graphically reflected broader economic inequalities – within limits.[18] Developing countries with large energy sectors, or energy-intensive manufacturing sectors, had higher emissions-per-head levels than those without, but this reflected only the state of their export industries, not the level of fossil fuel use or other emissions-producing activity by their citizens. Moreover, the research showed inequalities *between* nations, but not *within* nations.

In the 2000s, against a background of deepening global inequalities on one hand, and the sharp increase in CO_2 emissions from developing countries on the other, efforts were made to refine the quantitative research, including consumption-based accounting of CO_2 emissions, i.e. counting the emissions associated with goods and services in the country in which they were consumed, rather than the one in which they were produced.[19] So the emissions caused by the production of 1000 steel bars manufactured in China and exported to the USA would be attributed to the USA. The methodology could only imperfectly capture the effects of inequalities between nations: in the example given, it counted the emissions from the steel mill but not, for instance, of the urban infrastructure required to support it.

Another attempted refinement sought to gauge – again using 'top-down' statistical methods – inequalities in emissions within nations. A US-based research team sought to apply the principle of 'common but differentiated responsibilities' for CO_2 emissions (as laid down by the UNFCCC) to 'individuals instead of nations'. The conclusions – that 1.13 billion 'high emitters', including many in developing countries, would have to help meet 'tough global atmospheric stabilisation targets', and that 'no country gets a pass', because even the poorest countries have individuals who exceed the researchers' 'universal emission cap' – were criticised for being 'at odds with economic theory and empirical data'. The authors' assumption that they could derive the level of individuals' emissions from income data was 'implausible', the critics argued.[20] The research also failed to solve the wider problem with emissions-per-capita statistics, that they do not (and can not) reflect the way that fossil fuel consumption and other emissions-generating activities are carried out by social and economic systems rather than individuals.

An attempt by two high-profile economists to pursue the 'individuals instead of nations' approach,[21] in support of proposals for an 'equitable adaptation fund', points to a still more eye-watering gulf between the richest and poorest individuals, but has not rectified the problems inherent in methods based on consumption-per-head statistics. Recent research that quantified greenhouse gas emissions according to nations and economic sectors, while developing an updated form of net energy analysis to account for trade flows (e.g. manufacture of energy-intensive goods in one country for consumption in another country)[22] may open up a more promising avenue.

Measuring primary and secondary energy consumption

The building blocks for analyses such as those mentioned above, or for narratives such as the one in this book, are raw statistics that measure physical phenomena: population, volumes of CO_2 emissions, and so on. Statistical measurements of international primary and secondary energy consumption, including fossil fuel consumption, are published by the International Energy Agency (IEA); the UN Statistics Division; the Energy Information Administration (EIA) of the US Department of Energy; and BP, the oil company. For readers outside the university system, the BP statistics are most easily available; they are published annually in the company's *Statistical Review of World Energy*, and can be downloaded as PDF or EXCEL files. The Shift Project Data Portal is a reliable secondary online source of statistical series.[23] The IEA statistics, which start from 1971, give the most comprehensive sectoral breakdown, and the clearest view of the distinction between primary and secondary energy. Selections are available on open access on the IEA web site.

APPENDIX 1. MEASURING IMPACTS, FLOWS AND INEQUALITIES

There are striking differences between these sets of statistics, concerning (i) non-commodified energy, and (ii) renewable energy sources.

The *BP Statistical Review* and the EIA's international statistical series count only commercially traded energy, and thereby exclude the biomass that dominates the energy balance of many African and Asian countries. (This provides a telling insight into how BP executives, and the US government, see the world). The IEA and UN statistics, by contrast, include estimates of non-commercial energy use.

Energy from renewable sources (hydro, wind and solar) used to generate electricity is treated differently by different statisticians. The dilemma arises because the methodologies were developed at a time when attention was focused on the ratio between primary energy supplied to electricity generation and secondary energy produced as electricity. In the case of electricity generation from hydro, wind or solar, the primary energy is not transformed chemically (as fossil fuels are). These are flows, rather than stocks: a small proportion of the primary energy (such as, the sun's rays or the force of falling water) drives the technological process but is left chemically unchanged. Engineers' attention is focused on technical improvements that can raise the proportion of energy captured.

The statisticians' dilemma is: what number should go in the primary energy consumption columns to represent these renewable inputs? The IEA and UN Statistics Division use the physical energy content method: they count the energy content of the electricity produced by the dam, wind turbine or solar panel, and use that number to represent primary energy consumption (primary energy content is assumed to be the same as secondary energy content). BP and EIA statisticians, in accordance with a predominantly commercial approach, use the partial substitution method: they count the energy content of the electricity produced by the renewable source, calculate how much primary fossil energy would have been needed to produce that electricity in a 38-per-cent-efficient plant, and use that number to represent the primary energy.[24] So their numbers representing primary energy from renewables are more than two-and-a-half times higher than the IEA's. This makes a noticeable difference in countries where hydro makes a significant contribution to electricity generation. For example, hydro's contribution to Norway's energy balance was in 2012 measured as 12.2 million toe by the IEA (using the physical energy content method) and 32.3 million toe by BP (using the partial substitution method).

Appendix 2
Additional figures and tables

The following figures and tables provide additional context for the text. The chapters to which they are relevant are indicated.

For Chapter 1. See p. 20.

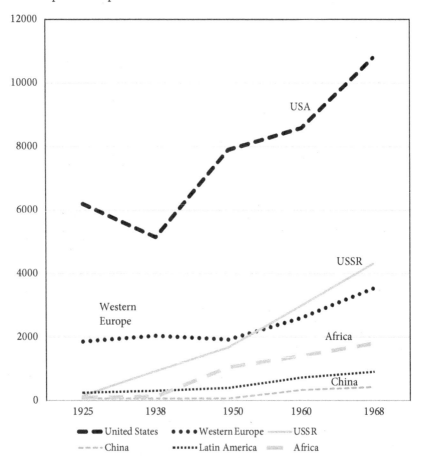

Figure 10 Commercial energy consumption per person per year, 1925–68
Kilogrammes of coal equivalent
Source: Darmstadter, *Energy in the World Economy*, p. 10

For Chapter 4. See p. 65 and p. 67.

Table 11 Nigeria energy balance, and uses of crude oil
Millions of tonnes of oil equivalent

		1971	1991	2011
Total primary energy supply		33.2	69.6	126.9
Coal & peat		0.1	0	0
Oil & oil products		1.7	11.2	11.8
Gas		0.2	4.0	12.2
Hydro & other renewables		31.2	54.4	102.8
Energy supply to final consumption		31.9	62.2	111.3
Energy supply for transformation		2.0	9.0	19.7
Coal & peat		0.3	0.1	0
Oil & oil products		0.2	3.3	0.7
Gas		0.1	3.3	8.8
Hydro & other renewables		1.0	1.5	8.2
Other/statistical adjustment		0.4	0.8	2.1
Energy sources for transformation (= sum of energy used and outputs)		2.0	9.0	19.7
Transformation: energy used				
Energy used in producing electricity		0.2	1.3	2.7
Energy used in oil refineries		0	0.2	0.1
Other energy industry own use and losses		0.2	2.8	5.2
Other transformation		0.9	1.0	7.7
Transfers & statistical differences		0.6	3.0	2.1
Transformation: outputs				
Electricity		0.1	0.7	2.0
Final consumption: energy sources		31.9	62.2	111.3
Coal & peat		0.1	0	0
Oil & oil products		1.5	7.9	11.2
Gas		0	0.7	3.5
Hydro & other rnwbls		30.2	52.9	94.6
Electricity		0.1	0.7	2.0
Final consumption: uses of energy		31.9	62.2	111.3
Industry	Iron & steel	0	0.1	0.3
	Chemical & petrochemical	0	0.3	0.2
	Other industry	0.8	1.8	9.3
Transport	Road	0.8	5.1	8.5
	Rail	0.1	0	0
	Domestic navigation and other transport	0.1	0	0
Residential		30.0	52.2	88.2
Commercial and public services		0	1.9	2.3
Other		0	0	1.5
Non-energy use	Chemical feedstocks	0	0	0.9
	Other non-energy use	0.1	0.7	0.1
Memo items				
Oil	*Produced and exported*	75.1	79.5	123.9
	Produced for domestic use	2.0	14.6	6.0

Notes: More than 99% of the 'hydro and other renewables' in Nigeria comprises biofuels and waste, i.e. fuelwood and other non-commercial fuels. This information is presented as a graphic, in Figure 7.
Source: IEA Energy Balances

Table 12 Global uses of electricity

	1971		1991		2011	
	mtoe	%	mtoe	%	mtoe	%
Total final consumption of electricity	377518	100%	860317	100%	1587297	100%
Iron and steel	28614	7.6%	44881	5.2%	91413	5.8%
Chemical & petrochemical industries	37928	10.0%	64877	7.5%	89183	5.6%
Non-ferrous metals	23816	6.3%	41990	4.9%	72738	74.6%
Non-metallic minerals	10279	2.7%	19852	2.3%	44492	2.8%
Other industry	103715	27.5%	216230	25.1%	375591	23.7%
Rail	9339	2.5%	8968	1.0%	18444	1.2%
Other transport	373	0.1%	11665	1.4%	6774	0.4%
Residential	85706	22.7%	227988	26.5%	428672	27.0%
Commercial and public services	55147	14.6%	181057	21.0%	363301	22.9%
Other	22601	6.0%	42809	5.0%	96690	6.1%

Source: IEA Energy Balances

For chapter 5. See p. 80.

Table 13 Shares of global oil production, per cent

per cent	1929	1950	1965	1972
North America	68.3	53.3	31.2	24.2
Europe	2.8	1.9	2.5	1.6
USSR	6.4	7	15.5	15.2
Africa	0.1	0.4	6.8	10.4
Latin America	16.1	19.2	14.4	8.9
Middle East	6.3	18.2	26.7	34.4
East Asia	n/a	n/a	2.9	5.3
Total	100	100	100	100
World total production, tonnes	211.1	537	1567.9	2640.6

Sources: UN, *World Energy Supplies in Selected Years, 1929-1950*; BP Statistical Review (for 1965 and 1972)

For chapter 6. See p. 99.

Table 14 Prices of petrol and electricity, 1980

	USA	West Germany	France	UK
Petrol price, US cents per litre				
Price for final consumer	31.42	62.3	75.7	64.2
Including tax	3.17	30.8	43.2	26.7
Average electricity price, US cents per kWh				
For industry	3.69	8	6.71	5.5
For households	5.36	12.27	11.02	8.11

Source: Claire Doblin, *The Growth of Energy Consumption and Prices*, pp. 46-9

For Chapter 10. See p. 154.

Table 15 Industry's role in fossil fuel consumption

In millions of tonnes of oil equivalent (total and by industry), and industry as % of total

	1980			1990			2000			2010		
	Total	By industry	*Ind'y as % of total*	Total	By industry	*Ind'y as % of total*	Total	By industry	*Ind'y as % of total*	Total	By industry	*Ind'y as % of total*
China	493.9	195.0	*39.5%*	664.2	238.9	*42.7%*	815.5	384.4	*47.1%*	1525.7	828.1	*54.3%*
India	173.8	47.2	*27.2%*	243.5	80.2	*32.9%*	315.4	110.3	*35.0%*	474.7	186.2	*39.2%*
Indonesia	49.6	8.0	*16.0%*	79.9	25.5	*31.9%*	121.2	40.4	*33.3%*	149.7	49.7	*33.2%*
Thailand	15.2	4.2	*27.7%*	28.9	9.1	*31.5%*	50.6	22.4	*44.2%*	84.9	44.6	*52.5%*
Non OECD, excluding the former												
Soviet Union	1487.0	532.4	*35.8%*	1983.5	745.9	*37.6%*	2566.7	991.8	*38.6%*	4020.7	1751.5	*43.6%*
US	1311.3	485.3	*37.0%*	1293.5	397.9	*30.8%*	1546.2	480.5	*31.1%*	1500.8	398.1	*26.5%*
UK	131.3	46.0	*35.0%*	138.2	42.6	*30.8%*	150.7	45.1	*29.9%*	137.4	32.5	*23.6%*
Germany	248.7	101.2	*40.7%*	240.8	88.6	*36.8%*	231.4	76.0	*32.8%*	229.0	77.6	*33.9%*
Italy	102.2	44.5	*43.6%*	114.9	44.2	*38.4%*	128.8	46.0	*35.7%*	130.0	39.4	*30.3%*
OECD	2941.6	1177.7	*40.0%*	3105.5	1107.8	*35.7%*	3642.6	1269.6	*34.9%*	3686.7	1152.2	*31.3%*

Note. For Germany, the 1980 figure is for the western part only. Source: IEA Energy Balances

Notes

Introduction

1. See pp. 9–10.
2. World consumption of commercially-produced coal, gas and oil was 7,166 million tonnes of oil equivalent (toe) in 1990 and 11,354.3 million toe in 2016, an increase of 58 per cent. *BP Statistical Review of World Energy*, 2017.
3. Vaclav Smil, *Energy in World History* (Oxford: Westview Press, 1994); V. Smil, *Energy at the Crossroads: global perspectives and uncertainties and other works* (London, MIT Press, 2005); Janet Ramage, *Energy: a guidebook* (Oxford: Oxford University Press, 1997); Bruce Podobnik, *Global Energy Shifts: fostering sustainability in a turbulent age* (Philadelphia: Temple University Press, 2006); Matthew Huber, 'Energizing historical materialism: fossil fuels, space and the capitalist mode of production', *Geoforum* 40 (2008), pp. 105–15; M. Huber, *Lifeblood: oil, freedom and the forces of capital* (Minneapolis: Quadrant, 2013); Timothy Mitchell, *Carbon Democracy: political power in the age of oil* (London: Verso, 2011); Andreas Malm, *Fossil Capital: the rise of steam power and the roots of global warming* (London: Verso, 2016).
4. David Nye, *Consuming Power* (London: MIT Press, 2001); D. Nye, *Electrifying America* (London: MIT Press, 1990); Sunila Kale, *Electrifying India: regional political economies of development* (Stanford: Stanford University Press, 2014).
5. Christophe Bonneuil and Jean-Baptiste Fressoz, *The Shock of the Anthropocene* (London: Verso, 2016), pp. 99–100; Frank Trentmann, *Empire of Things: how we became a world of consumers, from the fifteenth century to the twenty-first* (London: Allen Lane, 2016), pp. 676–90; Adam Tooze, 'A sweeping history', *Guardian*, 25 June 2016.
6. See pp. 38–40.
7. For examples and further explanation, see pp. 58–60.
8. The countries that joined the OECD in the 1960s and 1970s are: Australia, Austria, Belgium, Canada, Denmark, Finland, France, Germany, Greece, Ireland, Italy, Japan, Netherlands, New Zealand, Norway, Portugal, Spain, Sweden, Switzerland, Turkey, UK and the USA. In the 1990s and 2000s, Chile, Estonia, Hungary, Israel, Korea, Latvia, Mexico, Poland, Slovakia and Slovenia were added.

Chapter 1

1. The debates on the Anthropocene are reviewed in: Bonneuil and Fressoz, *The Shock*. See also: Will Steffen, A. Sanderson, P. Tyson et al., *Global Change and the Earth System: a planet under pressure* (New York: Springer-Verlag, 2004); W. Steffen, Paul Crutzen and John McNeill, 'The Anthropocene: are humans now overwhelming the great forces of nature?', *Ambio* 36 (December 2007), pp. 614–21; and Simon Lewis and Mark Maslin, 'Defining the Anthropocene', *Nature* 519 (March 2015), pp. 171–80. On the 'great acceleration', see: J. McNeill and Peter Engelke, *The Great Acceleration: an environmental history of the anthropocene since 1945* (Cambridge, Mass: The Belknap Press, 2014); Kathy Hibbard et al., 'Group Report: Decadal-scale interactions of humans and the environment', in Robert Costanza, Lisa Graumlich and Will Steffen (eds.), *Sustainability or Collapse? An integrated history and future of people on earth*

(London: MIT Press, 2011), pp. 341–75; and W. Steffen et al, 'The trajectory of the anthropocene: the Great Acceleration', *Anthropocene Review* 2/1 (April 2015), 81–98.
2. Alfred Crosby, *Children of the Sun: a history of humanity's unappeasable appetite for energy* (London: W. W. Norton, 2006), pp. 59–60 and 68–9; Podobnik, *Global Energy Shifts*, pp. 19–20; Paul Rutter and James Keirstead, 'A brief history and the possible future of urban energy systems', *Energy Policy* 50 (2012), pp. 72–80; Kenneth Pomeranz, *The Great Divergence: China, Europe and the making of the modern world economy* (Princeton: Princeton University Press, 2000), pp. 59–68.
3. Ellen Meiksins Wood, *The Origin of Capitalism* (New York: Monthly Review Press, 1999); Joel Mokyr, 'Introduction' in Mokyr (ed.), *The British Industrial Revolution: an economic perspective* (Oxford: Westview Press, 1999), pp. 7–10; Vaclav Smil, *Energy in World History*, pp. 193–4; Joseph Inikori, *Africans and the Industrial Revolution in England: a study in international trade and economic development* (Cambridge: Cambridge University Press, 2002), especially pp. 476–82.
4. David Landes, *The Unbound Prometheus: technological change and industrial development in Western Europe from 1750 to the present* (Cambridge: Cambridge University Press, 1969), pp. 86, 88–96 and 99–104; Robert Allen, *The British Industrial Revolution in Global Perspective* (Cambridge: Cambridge University Press, 2009), pp. 217–20; Lewis Mumford, *Technics and Civilization* [1934] (London: Routledge & Kegan Paul, 1955), pp. 153–62; Arnulf Grubler, *Technology and Global Change* (Cambridge: Cambridge University Press, 1998), p. 249.
5. Edward Wrigley, *Continuity, Chance and Change: the character of the Industrial Revolution in England* (Cambridge: Cambridge University Press, 1988), pp. 5–6, 19–30 and 95; E. Wrigley, *Energy and the English Industrial Revolution* (Cambridge: Cambridge University Press, 2010), especially pp. 16–17 and 39–46; Rolf Peter Sieferle, *The Subterranean Forest: energy systems and the industrial revolution* (Cambridge: The White Horse Press, 2001); Roger Fouquet, *Heat, Power and Light: revolutions in energy services* (Cheltenham: Edward Elgar, 2008), pp. 52–4. See also: Stefania Barca, 'Energy, property and the industrial revolution narrative', *Ecological Economics* 70 (2011), pp. 1309–15; Paolo Malanima, 'Energy crisis and growth 1650-1850: the European deviation in a comparative perspective', *Journal of Global History* 1 (2006), pp. 101–21.
6. Allen, *The British Industrial Revolution*, pp. 2–4; H. J. Habakkuk, *American and British Technology in the Nineteenth Century: the search for labour-saving inventions* (Cambridge: Cambridge University Press, 1962).
7. Arnulf Grubler, 'Grand Designs: historical patterns and future scenarios of energy technological change', in Arnulf Grubler and Charlie Wilson (eds.), *Energy Technology Innovation: learning from historical successes and failures* (Cambridge: Cambridge University Press, 2014), pp. 39–53, here pp. 43–4.
8. Allen, *The British Industrial Revolution*, pp. 171–80; Malm, *Fossil Capital*, pp. 121–222.
9. Landes, *The Unbound Prometheus*, p. 97; Robert Ayres, *Technological Transformations and Long Waves*. Report RR-89-1 (Laxenburg: IIASA, 1989), pp. 20–22.
10. Rutter and Keirstead, 'A brief history'; Smil, *Energy in World History*, p. 160; Nye, *Electrifying*, pp. 238–9.
11. Landes, *The Unbound Prometheus*, p. 99.
12. Podobnik, *Global Energy Shifts*, p. 29.
13. Podobnik, *Global Energy Shifts*, pp. 33–7.
14. In the 2010s, electricity production accounted for about 35 per cent, and transport about 24 per cent, of global fossil fuel use (IEA statistics). V. Smil, *Creating the Twentieth Century: technical innovations of 1867-1914 and their lasting impact* (Oxford: Oxford University Press, 2005), especially pp. 22–7.
15. Bouda Etemad and Jean Luciani, *World Energy Production/production mondiale de l'énergie 1800-1985* (Geneva: Droz, 1991), pp. 202, 207 and 223; Podobnik, *Global Energy Shifts*, pp. 35–7; Landes, *The Unbound Prometheus*, p. 293; Jean-Claude Debeir,

Jean Paul Deléage and Daniel Hémery, *In the Servitude of Power: Energy and Civilization Through the Ages* (London: Zed Books, 1991), pp. 106-7.

16. In a steam turbine, jets of steam hit the turbine's blades and produce rotary motion, a system into which Parsons introduced crucial efficiency improvements. Earlier steam engines were reciprocating engines in which the steam drove pistons.

17. Geoffrey Boyle, John Everett and Janet Ramage, *Energy Systems and Sustainability: power for a sustainable future* (Oxford: Oxford University Press, 2004), pp. 209-15 and 334-47; Landes, *The Unbound Prometheus* pp. 277-84; Richard Rudolph and Scott Ridley, *Power Struggle: the hundred-year war over electricity* (New York: Harper & Row, 1986), pp. 36-8.

18. Nye, *Electrifying*, pp. 167-70; William Hausman, Peter Hertner and Mira Wilkins, *Global Electrification: multinational enterprise and international finance in the history of light and power, 1878-2007* (Cambridge: Cambridge University Press, 2008), pp. 19-20; Thomas Hughes, *Networks of Power: Electrification in Western Society, 1880-1930* (Baltimore: Johns Hopkins Press, 1983), p. 227.

19. Robert Redlinger, Per Andersen and Poul Morthorst, *Wind Energy in the 21st Century: economics, policy, technology and the changing electricity industry* (Basingstoke: Palgrave, 2001), pp. 42-5; Robert Righter, *Wind Energy in America: a history* (London: University of Oklahoma Press, 1996), pp. 61-100.

20. Nye, *Electrifying*, pp. 138-40; Rudolph and Ridley, *Power Struggle*, pp. 36-41; Daniel Yergin, *The Quest: energy, security and the remaking of the modern world* (London: Penguin, 2012), pp. 354-5.

21. Ellen Leopold and David McDonald, 'Municipal Socialism Then and Now: some lessons for the Global South', *Third World Quarterly* 33:10 (2012), pp. 1837-53.

22. Hellmut Wollmann, Harald Baldersheim, Giulio Citroni, Gérard Marcou and John McEldowney, 'From public service to commodity: the demunicipalisation (or remunicipalisation?) of energy provision in Germany, Italy, France, the UK and Norway', in Wollman and Marcou (eds.), *The Provision of Public Services in Europe: between state, local government and market* (Cheltenham: Edward Elgar, 2010), pp. 168-90.

23. Nye, *Electrifying*, pp. 186-204; V. Smil, *Energy in Nature and Society: general energetics of complex systems* (London: MIT Press, 2008), p. 268; Warren Devine, 'From shafts to wires: historical perspective on electrification', *Journal of Economic History* 43:2 (1983), pp. 347-72; Frederik Nebeker, *Dawn of the Electronic Age: electrical technologies in the shaping of the modern world, 1914-1945* (Piscataway, NJ: IEEE, 2009), p. 93; Hausman et al, *Global Electrification*, p. 18.

24. Nye, *Consuming*, p. 166; Yergin, *The Quest*, pp. 352-3; Philip Schewe, *The Grid: a journey through the heart of our electrified world* (Washington: Joseph Henry Press, 2007), pp. 69-75.

25. Smil, *Energy in World History*, pp. 167-9 and 181-5; Landes, *The Unbound Prometheus*, pp. 280-81; Debeir et al, *In the Servitude*, pp. 124-5.

26. Refining separates crude oil, a mixture of various hydrocarbon chemicals, into different products, from the most solid such as bitumen and tar, to fuel oils, to the lightest liquids (petrol for cars) and gases such as propane and ethane. To break down heavier hydrocarbons and increase the yield of light ones, thermal cracking used heat; catalytic cracking uses chemical catalysts.

27. Joel Darmstadter, *Energy in the World Economy: a statistical review of trends in output, trade and consumption since 1925* (Baltimore: Johns Hopkins Press, 1971), p. 10 and pp. 16-17; Daniel Yergin, *The Prize: the epic quest for oil, money and power* (New York: Free Press, 2009), pp. 95-6, 151-61 and 365.

28. United Nations, *World Energy Supplies in Selected Years, 1929-1950* (Statistical Papers series 1 no. 1) (New York: UN, 1952), p. 10 and p. 38.

29. UN, *World Energy*, p. 6. The UN's estimate of commercial energy supply, 2512.3 million tce in 1949, compares with Darmstadter et al's estimate of 2610.9 million tce in 1950.
30. Landes, *The Unbound Prometheus*, pp. 307-23 and 431-9; Gail Cooper, *Air-Conditioning America: engineers and the controlled environment* (Baltimore: Johns Hopkins University Press, 1998), pp. 58-75; Jonathan Rees, *Refrigeration Nation: a history of ice, appliances and enterprise in America* (Baltimore: Johns Hopkins University Press, 2013), p. 164.
31. Ronald Kline, *Consumers in the country: technology and social change in rural America* (Baltimore: Johns Hopkins University Press, 2000), p. 5; Nye, *Electrifying*, p. 167; M. Lapping, 'Rural policy and planning', *Rural Development Paper* 31 (Pennsylvania: Pennsylvania State University, 2005), p. 6.
32. Insull was arrested in Turkey in 1934 and, after repatriation, tried and acquitted. Rudolph and Ridley, *Power Struggle*, p. 64.
33. Hausman et al, *Global Electrification*, p. 197; Rudolph and Ridley, *Power Struggle*, p. 69 and pp. 76-7; Righter, *Wind Energy*, pp. 123-4.
34. Stan Luger, *Corporate Power, American Democracy and the Automobile Industry* (Cambridge: Cambridge University Press, 2000), p. 8 and p. 40; Nye, *Consuming*, pp. 178-80; Boyle et al, *Energy Systems*, pp. 302-3; Paul Baran and Paul Sweezy, *Monopoly Capital: an essay on the American economic and social order* (London: Penguin, 1966), p. 294.
35. Stephen Goddard, *Getting There: The Epic Struggle Between Road and Rail in the American Century* (Chicago: University of Chicago Press, 1994), pp. 102-17; UN Centre on Transnational Corporations, *Climate Change and Transnational Corporations* (New York: UN, 1992), p. 58.
36. David St Clair, *The Motorization of American Cities* (New York: Praeger, 1986), pp. 7-12; Luger, *Corporate Power*, p. 13; Paul Barrett, *The Automobile and Urban Transit* (Philadelphia: Temple University Press: 1983); Barry Commoner, *The Poverty of Power*, (London: Jonathan Cape, 1976) pp. 188-95; Bradford C. Snell, 'American Ground Transportation: a proposal for restructuring the industries' (Washington: US Government Printing Office, 1974); Nye, *Electrifying*, pp. 167-70.
37. Paul Hoffman, writing in the *Saturday Evening Post* in 1939, cited by St Clair, *The Motorization*, pp. 125-8.
38. Nye, *Consuming*, pp. 179-80.
39. Giles Slade, *Made to Break: technology and obsolescence in America* (Cambridge: Harvard University Press, 2006), pp. 4-5 and 74-6; Tim Cooper (ed.), *Longer Lasting Products: alternatives to the throwaway society* (Farnham: Gower, 2010), p. 11 and p. 221; David Nye, *Consuming*, pp. 175-6.
40. Alan Milward, *War, Economy and Society 1939-1945* (Berkeley: University of California Press, 1979), pp. 56-9 and 63-9; Michael Renner, 'Assessing the Military's War on the Environment', in: Lester Brown (ed.), *State of the World 1991* (New York: W.W. Norton, 1991), pp. 132-52; Smil, *Energy in Nature*, p. 354; John Miller, *Men and Volts at War: the story of General Electric in World War II* (New York: McGraw Hill, 1947), p. 176; Adam Tooze, *The Wages of Destruction: the making and breaking of the Nazi economy* (London: Penguin, 2006), pp. 411-25 and 493-511; Alec Nove, *An Economic History of the USSR* (London: Pelican, 1989), pp. 247-8 and 260-79; John Barber and Mark Harrison, *The Soviet Home Front* (London: Longman, 1991), pp. 133, 166-7 and 185-6.
41. Milward, *War*, pp. 197-8 and 245-346; Bonneuil and Fressoz, *The Shock*, pp. 138-47; John McNeill and David Painter, 'The global environmental footprint of the US military 1789-2003' in Charles Closmann (ed.), *War and the Environment: military destruction in the modern age* (College Station: Texas A&M University Press, 2009), pp. 10-31;

here p. 20; Alejandro Piera, 'Why taxes are not an option', www.greenaironline.com, 26 March 2015.
42. Michael Pollan, *The Omnivore's Dilemma: the search for a perfect meal in a fast-food world* (London: Bloomsbury, 2006), pp. 41–5.

Chapter 2

1. Definitions adapted from UNDP, *World Energy Assessment: energy and the challenge of sustainability* (New York: UNDP, 2000), pp. 175–6, and Thomas Johannson, Anand Patwardhan, Nebojsa Nakicenovic and Luis Gomez-Echeverri (eds.), *Global Energy Assessment* (Cambridge: IIASA/Cambridge University Press, 2012), p. 104. www.globalenergyassessment.org. In the latter, a separate step, 'secondary energy', is inserted between conversion and distribution of primary energy. See also Jeanne Anderer, Alan McDonald and Nebosja Nakicenovic, *Energy in a finite world: paths to a sustainable future* (Laxenburg: IIASA, 1981), pp. 3–4, and Daniel Spreng, *Net-Energy analysis and the energy requirements of energy systems* (New York: Praeger, 1988), pp. 6–7, 47–8 and 71.
2. UNDP, *World Energy Assessment*, p. 175.
3. W. Devine, 'Energy accounting for solar and alternative energy sources', in T. Nejat Veziroglu (ed.), *Alternative Energy Sources* II. Volume 9 (London: Hemisphere, 1981), pp. 3815–44. See also Appendix 1.
4. Amory Lovins, *The Energy Controversy: Soft Path Questions and Answers* (San Francisco: Friends of the Earth, 1979), pp. 15–34.
5. R. Ayres, *Energy Inefficiency in the US Economy: a new case for conservation* (Laxenburg: IIASA, 1989).
6. A. Lovins, *Soft Energy Paths: towards a durable peace* (Harmondsworth: Penguin, 1977), p. 40.
7. Smil, *Energy in Nature*, pp. 217–18.
8. Roberto Aguilera and Marian Radetzki, *The Price of Oil* (Cambridge: Cambridge University Press, 2016), pp. 37–40; Peter Berck and Michael Roberts, 'History of prices of energy', in C. Cleveland (ed.), *Concise Encyclopedia of History of Energy* (Oxford, Elsevier, 2009), pp. 241–9.
9. Smil, *Energy in Nature*, p. 266. Author's calculations from IEA energy balances.
10. The relevant thermodynamic principles are summarised in Nicolas Carnot's theorem, published in 1824, which specifies the maximum efficiency a heat engine can obtain.
11. Boyle, Everett and Ramage, *Energy Systems*, pp. 221–2 and 368; Smil, *Energy at the Crossroads*, pp. 41–2; Roger Revelle, 'Energy Use in Rural India', in Vaclav Smil and William Knowland (eds.), *Energy in the developing world: the real energy crisis*: (Oxford: Oxford University Press, 1980), pp. 194–207, here p. 201; IEA, *Worldwide trends in energy use and efficiency: key insights* (Paris: IEA, 2008), pp. 71–4. See also: Robert Smith and Iain MacGill, 'Revolution, evolution or back to the future? Lessons from the electricity supply industry's formative days', in Fereidoon Sioshansi (ed.), *Distributed Generation and its Implications for the Utility Industry* (Amsterdam: Academic Press 2014), pp. 475–500. They note that electricity networks today are only 'modestly better in terms of cost and efficiency' than those of 40 years ago.
12. Vaclav Smil, *Two Prime Movers of Globalisation: the history and impact of diesel engines and gas turbines* (London: The MIT Press, 2010), pp. 150–51; Rolf Kehlhofer, *Combined Cycle Gas and Steam Turbine Power Plants* (Tulsa: Penn Well Publishing, 1997); Walter Patterson, *Transforming electricity: the coming generation of change* (London: Royal Institute for International Affairs, 1999), p. 74.
13. IEA, *Tracking Clean Energy Progress 2015* (Paris: IEA, 2015), p. 52; Boyle, Everett and Ramage, *Energy Systems*, p. 370.

14. Leslie Dienes and Theodore Shabad, *The Soviet Energy System: resource use and policies* (Washington: V.H. Winston, 1979), pp. 23 and 197; C. McDonald, 'An evaluation of the potential for district heating in the US', T. J. Verizoglu (ed.), *Alternative Energy Sources: an international compendium*, vol. 9 (London: Hemisphere, 1978), pp. 4107–30; Euroheat & Power, *District Heating and Cooling Survey* (Brussels: EHP, 2009), p. 392; John Beldin, 'Conservation as an energy source', in Verizoglu (ed.), *Alternative Energy*, vol. 9, pp. 4279–91; Benjamin Sovacool, 'How long will it take? Conceptualising the temporal dynamics of energy transitions', *Energy Research & Social Science* 13 (2016), pp. 202–15; Marianne van der Steen et al, 'Evolutionary innovation systems of low carbon electricity', in Timothy Foxon et al (eds.), *Innovation for a Low Carbon Economy: economic, institutional and management approaches* (Cheltenham: Edward Elgar, 2008), pp. 175–202.
15. Jose Goldemberg, Thomas Johannson, Amulya Reddy and Robert Williams, *Energy for a sustainable world* (New York: Wiley, 1988), p. 151; A. Lovins, 'Energy Strategy: the Road Not Taken', *Foreign Affairs* 55 (1976), pp. 65–96; David Morris, 'The Pendulum Swings Again: a century of urban electric systems' in H. Brown (ed.), *Decentralising Electricity Production* (London: Yale University Press, 1983), pp. 37–58; Simon Pirani, *Elusive Potential: Natural Gas Consumption in the CIS and the question for efficiency* (Oxford: OIES, 2011), pp. 36–8 and 56–70; A.S. Nekrasov et al, 'Sovremennoe sostoianiie teplosnabzheniia Rossii', *Problemy prognozirovaniia* no.1 (2011), pp. 30–43.
16. A. Lovins, *Small is Profitable* (Snowmass, Col.: Rocky Mountain Institute, 2002), p. 44; Lovins, *The Energy Controversy*, p. 24.
17. W. Patterson, *Keeping the Lights On: towards sustainable electricity* (London: Earthscan, 2007), pp. 115–21.
18. David Nye, 'Consumption of Energy', in Trentmann (ed.), *Oxford Handbook of the History of Consumption* (Oxford: Oxford University Press, 2012), pp. 307–25, here p. 317; Gijs Mom, *The Electric Vehicle: technology and expectations in the automobile age* (London: Johns Hopkins University Press, 2004), pp. 275–301; David Kirsch, *The Electric Vehicle and the Burden of History* (London: Rutgers University Press, 2000); Gregory Nemet, 'Automobile fuel efficiency standards', in Arnulf Grubler and Charlie Wilson, *Energy Technology Innovation: learning from historical successes and failures* (Cambridge University Press, 2014), pp. 178–92.
19. Smil, *Energy in Nature*, p. 266.
20. Smil, *Energy at the Crossroads*, p. 13; Clayton Christensen, *The Innovators' Dilemma: when new technologies cause great firms to fail* (Boston: Harvard Business School Press, 1997), pp. 89–91.
21. Bertrand Chateau and Bruno Lapillonne, *Energy demand: facts and trends. A comparative analysis of industrialised countries* (Vienna: Springer-Verlag, 1982), p. 174; IEA, *Worldwide Trends*, p. 34.
22. Julian Allwood, Jonathan Cullen et al, *Sustainable Materials With Both Eyes Open* (Cambridge: UIT, 2012), p. 306.
23. V. Smil, *Enriching the Earth: Fritz Haber, Carl Bosch and the Transformation of Food Production* (London: MIT Press 2001), p. 244; G. J. Leigh, *The World's Greatest Fix: a history of nitrogen and agriculture* (Oxford: Oxford University Press, 2004), pp. 140–54.
24. Albert Bartlett, 'Forgotten Fundamentals of the Energy Crisis', *American Journal of Physics* 46:9 (1978), pp. 876–88; Huber, *Lifeblood*, p. 87.
25. R. Stephen Berry and Margaret Fels, 'The Energy Cost of Automobiles', *Bulletin of the Atomic Scientists* 10, December 1973, pp. 11–17 and 58–60; Inter Technology Corporation, *Analysis of the economic potential of solar thermal energy to provide industrial process heat* (San Mateo: Solar Energy Information Services, 1977); Spreng, *Net-Energy Analysis*, pp. 61–2.
26. Amory Lovins, *Energy-efficient buildings: institutional barriers and opportunities* (Boulder: E Source, 1992).

27. John Cullen, Julian Allwood and Margarita D. Bambach, 'Mapping the global flow of steel: from steelmaking to end-use goods', *Environmental Science and Technology* 2012 (46), pp. 13048–55; Allwood, Cullen et al, *Sustainable Materials*, pp. 213–14; J. Cullen and J. Allwood, 'Mapping the global flow of aluminium: from liquid aluminium to end-use goods', *Environmental Science and Technology* 2013 (47), pp. 3057–64.
28. Allwood, Cullen et al, *Sustainable Materials*, pp. 328–9.
29. Jonathan Cullen, Julian Allwood and Edward Borgstein, 'Reducing energy demand: what are the practical limits?', *Environmental Science and Technology* 45 (2011), pp. 1711–18.
30. Hermann Scheer, *Energy Autonomy: the economic, social and technological case for renewable energy* (London: Earthscan, 2007), p. 104.
31. Concentrator photovoltaics are distinct from concentrated solar power (CSP) systems that use mirrors and lenses to concentrate sunlight, and use the heat produced to generate electricity, usually with a steam turbine.
32. John Perlin, *The Silicon Solar Cell Turns 50*, https://nrel.gov, 2004; 'Sunny uplands', *Economist*, 21 November 2012; Keith Barnham, *The Burning Answer: a user's guide to the solar revolution* (London: Weidenfeld and Nicolson, 2014), pp. 124 and 132–3; Morgan Bazilian et al, 'Re-considering the Economics of Photovoltaic Power', *Renewable Energy* 53 (2012), pp. 329–38.
33. Gregory Nemet, 'Solar water heater innovation in the US, China and Europe', in Grubler and Wilson, *Energy Technology*, pp. 105–17; Ethan Kapstein, 'Transition to Solar Energy: An Historical Approach' in Lewis Perelman et al (eds.), *Energy Transitions: long-term perspectives* (Boulder, Colorado: Westview Press/AAAS, 1981), pp. 109–23.
34. Righter, *Wind Energy*, pp. 198–216; Lena Nij and Per D. Andersen, 'A comparative assessment of wind turbine innovation and diffusion policies', in Grubler and Wilson, *Energy Technology*, pp. 221–30.
35. Bernadett Kiss, Lena Neij and Martin Jakob, 'Heat pumps: a comparative assessment' in Grubler and Wilson, *Energy Technology Innovation*, pp. 118–32.
36. Timothy Gutowski et al, 'Thermodynamic analysis of resources used in manufacturing processes', *Environmental Science & Technology* 43 (2009), pp. 1584–90; V. Smil, *Making the Modern World: materials and dematerialisation* (Chichester: John Wiley, 2014), pp. 72–4 and p. 187; Bob Hughes, *The Bleeding Edge: why technology turns toxic in an unequal world* (Oxford: New Internationalist, 2016), pp. 190–203.
37. Electric Power Research Institute, *Estimating the Costs and Benefits of the Smart Grid* (Palo Alto: EPRI, 2011), section 4, p. 3; Johansson et al, *Global Energy*, pp. 1159–61.
38. Smith and MacGill, 'Revolution, evolution'.

Chapter 3

1. In 1980, the US government's National Research Council appointed a Committee on the Behavioural and Social Aspects of Energy Consumption and Production. It concluded that there were 'four different views of energy' widely held in US society: *as a commodity* (a view they described as 'dominant'); *as a social necessity to 'meet essential needs'* (close to my idea of a public service); *as a 'strategic material'*; and *as an ecological resource for humanity at large*. Paul Stern and Elliot Aronson (eds.), *Energy Use: the human dimension* (New York: W.H. Freeman & co., 1984), pp. 15–26.
2. David Nye, 'Consumption of Energy', pp. 310–11. See also: Larry Lohmann and Nick Hildyard, *Energy, Work and Finance* (Sturminster Newton: Cornerhouse, 2014).
3. Stephen Healy and Iain MacGill, 'From Smart Grid to Smart Energy Use', in F. Sioshansi (ed.), *Smart Grid: integrating renewable, distributed and efficient energy* (Oxford: Elsevier, 2012) pp. 29–59, here p. 30.

4. Author's estimates; UN, *World Energy Supplies*, pp. 10-11 and *BP Statistical Review of World Energy*.
5. Ronald Kline, *Steinmetz: engineer and socialist* (Baltimore: Johns Hopkins University Press, 1992), pp. 216-22; August Bebel, *Woman and Socialism* [1879], Chapter XXVII.3 and 4, www.marxists.org; Petr Kropotkin, *Fields, Factories and Workshops* [preface to second edition, 1912] (Montreal: Black Rose Books, 1994), pp. xx-xxi.
6. Emile Zola, *Work*, trans. Ernest Alfred Vizetelly (London: Chatto & Windus, 1901), pp. 367-8.
7. Sarmila Bose, *Money, Energy and Welfare: the state and the household in India's rural electrification policy* (Delhi: Oxford University Press, 1993), pp. 10-11.
8. Elizabeth Cecelski, Joy Dunkerley and William Ramsay, *Household energy and the poor in the third world* (Washington: Resources for the Future, 1979), pp. 9-10; Shonali Pachauri and Leiwen Jiang, *The Household Energy Transition in India and China* (Laxenburg: IIASA, 2008), pp. 17-18; Bose, *Money, Energy and Welfare*, p. 31.
9. Johansson et al, *Global Energy*, pp. 1340-43.
10. Aggregate commercial energy consumption by industry in OECD countries was 20 per cent lower in 1983 than in 1973. IEA, *Thirty Years of Energy Use in IEA Countries: oil crises and climate challenges* (Paris: IEA/OECD, 2004), pp. 65-6.
11. L. Bernstein and J. Roy, 'Industry', in IPCC, *Climate Change 2007: Mitigation. Contribution of Working Group III to the Fourth Assessment Report of the IPCC* (Cambridge: Cambridge University Press, 2007), pp. 451-2.
12. Total final energy consumption by industry (including consumption of fossil fuels directly, and electricity and heat consumption, but not including gas and coal used as feedstocks in chemical processing) rose from 1.4 billion toe to 2.5 billion toe (IEA Energy Balances); Goldemberg et al, *Energy for a sustainable world*, p. 20.
13. J. Allwood et al., 'Industry 1.61803: the transition to an industry with reduced material demand fit for a low carbon future', *Philosophical Transactions of the Royal Society A* vol. 375, no. 2095 (13 June 2017) p. 20160361; Jennifer Clapp, 'The distancing of waste: overconsumption in a global economy', pp. 155-76, in Thomas Princen, Michael Maniates and Ken Conca, *Confronting Consumption* (London: MIT Press, 2002).
14. Anderer et al, *Energy in a finite world*, vol. 2, p. 16; UN Habitat, *State of the World's Cities 2008/9: Harmonious Cities* (UN Habitat/Earthscan, 2008), p. 133; Christopher Kennedy et al, 'Energy and material flows of megacities', *PNAS* vol. 112, no. 19 (2015), pp. 5985-90.
15. Johansson et al., *Global Energy*, p. 1310; Arnalf Grubler, 'Transitions in Energy Use', in Cleveland, *Concise Encyclopedia*, pp. 287-300. Cities' share of fossil fuel consumption should not be confused with their share of greenhouse gas emissions, which is significantly lower (because of the contribution of forest clearance, agriculture and so on). The widely repeated assertion that cities account for 80 per cent of emissions is questioned in: David Satterthwaite, 'Cities' contribution to global warming: notes on the allocation of greenhouse gas emissions', *Environment & Urbanisation* 20(2) (2008), pp. 539-49.
16. UN Habitat, *World Cities Report 2016: Urbanisation and Development. Emerging Issues* (Nairobi: UN, 2016), pp. 6-7; Johansson et al, *Global Energy*, pp. 1317-18.
17. Mike Davis, *Planet of Slums* (London: Verso, 2006), pp. 20-49; Patricia Romero Lankao, 'Are we missing the point? Particularities of urbanisation, substainability and carbon emissions in Latin American cities', *Environment and Urbanization* 19(1) (2007), pp. 159-75; UN Habitat, *The Challenge of Slums: global report on human settlements 2003* (London: Earthscan, 2003), p. 14.
18. UN, *World Cities Report 2016*, p. 92.
19. Johansson et al, *Global Energy*, pp. 583-4; IEA, *Thirty Years*, pp. 130-31.
20. Peter Newman and Jeffrey Kenworthy, *The End of Automobile Dependence: how cities are moving beyond car-based planning* (London: Island Press, 2015), p. 4 and p. 35.

21. Norman Myers with Jennifer Kent, *Perverse subsidies: tax $s undercutting our economies and environments alike* (Winnipeg: International Institute for Sustainable Development, 1998), p. xvii; 'Sacred spaces', *Economist*, 8 April 2017, pp. 20–22.
22. Benoit Lefevre, 'Urban Transport Energy Consumption: determinants and strategies for its reduction. An analysis of the literature', *SAPIENS journal* 2:3 (2009); Alain Bertaud, *Clearing the Air in Atlanta: transit and smart growth or conventional economics?*, 2002, www.alainbertaud.com.
23. Edward Glaeser and Matthew Kahn, 'Sprawl and Urban Growth' in J.V. Henderson and J. Thisse (eds.), *Handbook of Regional and Urban Economics*, vol. 4 (Amsterdam: Elsevier, 2004), pp. 2481–2527; UN, *World Cities Report 2016*, p. 5.
24. The HOMES project in the Netherlands distinguished between goods for 'modernisation' (substituting for domestic labour) and 'enrichment'. See B. Gatersleben and Ch. Vlek, 'Household Consumption, Quality of Life and Environmental Impacts: a psychological perspective and empirical study' in Jan Noorman Klaas and Ton Schoot Uiterkamp (eds.), *Green households? Domestic consumers, environment and sustainability* (London: Earthscan, 1998), pp. 141–83, here pp. 143–4.
25. Stern and Aronson (eds.), *Energy Use*, pp. 48–52.
26. David Goldblatt, *Sustainable energy consumption and society: personal, technological or social change* (Dordrecht: Springer, 2005), pp. 16–19; D. Goldblatt, 'A dynamic structuration approach to information for end-user energy conservation', *European Council for Energy Efficient Economy Summer Study proceedings*, 2003, pp. 1111–21.
27. Author's calculation from IEA Energy Balances and UN population statistics.
28. David Satterthwaite, 'The Implications of Population Growth and Urbanisation for Climate Change', *Environment and Urbanisation* 21:2 (2009), pp. 545–67; Thomas Princen, Michael Maniates and Ken Conca, 'Confronting Consumption', in Princen et al. (eds.), *Confronting Consumption* (London: MIT Press, 2002), pp. 1–20.
29. Shonali Pachauri, 'Household electricity access a trivial contributor to CO_2 emissions growth in India', *Nature Climate Change* 4 (2014), pp. 1073–6.
30. S. Pachauri, *An Energy Analysis of Household Consumption: changing patterns of direct and indirect use in India* (Dordrecht: Springer, 2004), p. 24; Darmstadter, *Energy in the World Economy*, pp. 9–12; Goldemberg et al, *Energy for a sustainable world*, p. 191; Smil, *Energy in Nature and Society*, p. 258; B. Podobnik, 'Global energy inequalities: exploring the long-term implications', *Journal of World-Systems Research*, VIII:2 (2002), pp. 252–74.
31. Pachauri, *An Energy Analysis*, pp. 164–7; Lucas Chancel and Thomas Piketty, *Carbon and Inequality: from Kyoto to Paris* (Paris: Paris School of Economics, 2015).
32. IEA, *World Energy Outlook 2006, 2007, 2010 and 2015*; Douglas Barnes, 'Two Billion People Gain Electricity Access 1970–2010', www.energyfordevelopment.com, 2 November 2014.
33. Brenda Boardman, *Fuel Poverty: from cold homes to affordable warmth* (London: Belhaven Press, 1991); B. Boardman, *Fixing fuel poverty: challenges and solutions* (London: Earthscan, 2010).
34. Sanchez, *The Hidden Energy Crisis* (Rugby: Practical Action Publishing, 2010), p. 13; V. Santhakumar, *Analysing Social Opposition to Reforms* (London: Sage, 2008) p. 115; Johansson et al, *Global Energy*, p. 1154.

Chapter 4

1. www.data.worldbank.org.
2. Steven Davis and Ken Caldeira, 'Consumption-based accounting of CO_2 emissions', *PNAS* 12 (107) (March 2010), pp. 5687–92. See also pp. 205–6.

3. *CDP Carbon Majors Report 2017* (CDP/Climate Accountability Institute, 2017); Richard Heede, 'Tracing anthropogenic carbon dioxide and methane emissions to fossil fuel and cement producers, 1854–2010', *Climatic Change* (2014) 122, pp. 229–41; Simon Pirani, 'Tackling the corporate power in fossil fuel production and use', London Green Left blog (22 August 2017), www.londongreenleft.blogspot.co.uk/2017/08/tracking-corporate-power-in-fossil-fuel.html.
4. Mark Thurber and Richard Morse (eds.), *The Global Coal Market: supplying the major fuel for emerging economies* (Cambridge: Cambridge University Press, 2015), p. 449.
5. The BP statisticians' method gives greater weight to the contributions of hydro and solar than the IEA statisticians' method. (See also pp. 206–7.) In 2014 the IEA counted hydro's share of world primary energy as 2.4 per cent and 'other' (including wind and solar) as 1.4 per cent. It counted 10.3 per cent from biofuels and waste, largely the non-commercial fuels that BP excludes. IEA, *Key World Energy Statistics* 2016, p. 6.
6. For an explanation of the science to a general readership, see the [US] National Academy of Sciences and the Royal Society, *Climate Change: evidence and causes. An overview*, www.nas-sites.org/americasclimatechoices (New York/London: NAS/RS, 2014).
7. UNFCCC, The Paris Agreement, http://unfccc.int/paris_agreement/items/9485.php. Petra Tschakert, '1.5°C or 2°C: a conduit's view from the science-policy interface at COP20 in Lima, Peru'. *Climate Change Responses* 2:3 (2015); James Hansen et al, 'Assessing "Dangerous Climate Change": required reduction of carbon emissions to protect young people, future generations and nature', *PLOS One*, 3 December 2013.
8. IPCC, *Climate Change 2014: Synthesis Report. Contribution of Working Groups I, II and III to the Fifth Assessment Report of the Intergovernmental Panel on Climate Change* (Geneva: IPCC, 2014), pp. 63–4. Note that any amount of carbon dioxide is 3.664 times heavier than the carbon contained within it, so the budget equates to 873 billion tonnes of carbon. For background, see the Global Carbon Project, www.globalcarbonproject.org.
9. IPCC, *Revised IPCC Guidelines for National Greenhouse Gas Inventories: Workbook* (Geneva: IPCC, 2006), p. 1.6; IEA, CO_2 *Emissions from Fuel Combustion Highlights* (Paris: IEA, 2016), pp. 23–6 and 149–52. On methane, Ken Costello, *Answering Questions about Methane Emissions from the Natural Gas Sector* (Silver Spring: National Regulatory Research Institute, 2015), p. 17; Chris Le Fevre, *Methane Emissions: from blind spot to spotlight* (Oxford: OIES, 2017), pp. 10–24.
10. Corinne Le Quere et al, 'Global Carbon Budget 2016', *Earth Systems Science Data* 8 (2016), pp. 605–49. Other teams of researchers reached similar results. BP's estimates for 2015 are that 13.147 billion toe of fossil fuels were used, producing 33.508 billion tonnes of carbon dioxide, *BP Statistical Review*, 2016.
11. 'Resources' are total physical quantities; 'reserves' refers to quantities whose location has been discovered and are deemed to be economically recoverable.
12. Malte Meinshausen et al, 'Greenhouse-gas emission targets for limiting global warming to 2°C', *Nature*, 30 April 2009, pp. 1158–63; Michael Jakob and Jerome Hilaire, 'Unburnable fossil-fuel reserves', *Nature* 517 (2015), pp. 150–52; Carbon Tracker Initiative/Grantham Research Institute, *Unburnable Carbon 2013: wasted capital and stranded assets* (London: Carbon Tracker, 2013), p. 4. See also IEA, *Energy and Climate Change: WEO Special Report* (Paris: IEA, 2015), p. 38.
13. David Murphy, 'The implications of the declining energy return on investment of oil production', *Phil Trans R Soc* A372 (2013), 20130126. See also Richard Heinberg and David Fridley, *Our Renewable Future: laying the path for one hundred percent clean energy* (London: Island Press, 2016), pp. 18–20; Charles Hall et al, 'EROI of different fuels and the implications for society', *Energy Policy* 64 (2014), pp. 141–52.

14. The IEA energy balances summarise primary energy sources, their transformation into final energy sources and their consumption, sector-by-sector. www.iea.org/statistics.
15. IEA, *Energy Statistics Manual* (Paris: IEA, 2005), p. 15; IEA, *Electricity Information 2015* (Paris: IEA, 2015), p. II.14; ABB, *The state of global energy efficiency* (Zurich: ABB, 2013), p. 9.
16. Author's calculations from BP statistics; UNDP *World Energy Assessment*, pp. 33–4.
17. Chateau and Lapillonne, *Energy demand*, pp. 135–7 (on the post war boom); IEA, *Thirty Years*, p. 67; Tables 5, 6 and 7.
18. L. Bernstein and J. Roy (eds.), 'Industry', in IPCC, *Climate Change 2007*, pp. 451–2. In Table 5-b, the rows 'iron and steel', 'chemical & petrochemical', 'non-ferrous metals' and 'cement and other non-metallic minerals', added together, comprised 36 per cent of aggregate industrial and construction use in 1971 and 49.8 per cent in 2011. There are amounts in other rows (e.g. 'blast furnaces' and 'coke ovens' in Table 5-a) used for materials manufacture, so the overall proportion is even greater. See Allwood et al, *Sustainable Materials*, p. 18.
19. *Handbook of World Steel Statistics 1978*, p. 1; *Steel Statistical Yearbook 2000*, p. 12; *Steel Statistical Yearbook 2015*, p. 2 (on world steel production).
20. Allwood et al, *Sustainable materials*, p. 35; Chateau and Lapillonne, *Energy demand: facts and trends*, p. 166.
21. Allwood, Cullen et al, *Sustainable Materials*, p. 78.
22. Allwood, Cullen et al, *Sustainable Materials*, p. 291; IEA, *Worldwide Trends*, p. 34; Johansson et al, *Global Energy*, pp. 525–6.
23. World Resources Institute, *Navigating the Numbers: greenhouse gas data and international climate policy* (Washington: WRI, 2005), pp. 72–3; Adam Hanieh, *Capitalism and Class in the Gulf Arab States* (New York: Palgrave Macmillan, 2011), pp. 181–2; Macquarie Commodities Research, *Global Petrochemical Industry Outlook* (conference presentation, 2013); Johansson et al, *Global Energy*, pp. 523–4.
24. V. Smil, *Making the Modern World*, pp. 40–42 and 62–3.
25. PlasticsEurope, *Insight into consumption and recovery in Western Europe* (Brussels: PlasticsEurope, 2001); *The New Plastics Economy: rethinking the future of plastics* (Ellen Macarthur Foundation, 2014), p. 24; Allwood, Cullen et al, *Sustainable Materials*, pp. 310–11.
26. Smil, *Energy in Nature and Society*, p. 354; Renner, 'Assessing the Military's War', in Lester Brown et al, *State of the World 1991*, pp. 132–52.
27. Sohbet Karbuz, 'US Military Energy Consumption: Facts and Figures', www.resilience.org 21 May 2007; Renner, 'Assessing the Military's War'.
28. S. Karbuz, 'US Military Energy Consumption', and 'How Much Energy Does the US military Consume? An Update' 5 August 2013, www.karbuz.blogspot.co.uk/2013/08/how-much-energy-does-us-military.html; Renner, 'Assessing the Military's War'; Barry Sanders, *The Green Zone: the environmental costs of militarism* (Oakland: AK Press, 2009), pp. 49–52.
29. The row 'other' in Table 5-b is the sum of the rows 'non-specified' (other) and 'non-energy use in other' in the IEA balances. For IEA directions to national statistical agencies, see www.iea.org/statistics/resources/balancedefinitions.
30. Table 5-b; Johansson et al, *Global Energy*, p. 585; Anthony Downs, *Still Stuck in Traffic: coping with peak-hour traffic congestion* (Washington: Brookings Institution, 2004), pp. 277–82; Feng An et al, *Global Overview of Fuel Efficiency and Motor Vehicle Emission Standards* (New York: UNDESA, 2011), p.1.
31. Johansson et al, *Global Energy*, p. 586 and p. 589; Spreng, *Net-energy systems*, p. 230.
32. International Civil Aviation Organisation, Facts and Figures, www.icao.int/sustainability/Pages/FactsFigures.aspx; Johansson et al, *Global Energy*, pp. 589–91;

Alice Bows-Larkin, 'All adrift: aviation, shipping and climate change policy', *Climate Policy* 15:6 (2015), pp. 681–702.
33. Deborah Kramer, *Mineral Commodity Profiles: Nitrogen* (Washington: US Geological Survey, 2004), p. 43; Smil, *Enriching*, pp. 109–11; Tony Weis, *The Ecological Hoofprint: the global burden of industrial livestock* (London: Zed Books, 2013), p. 106.
34. GRAIN, *Food and climate change: the forgotten link* (Barcelona: GRAIN, 2011).
35. Roberto Sainz, *Framework for calculating fossil fuel use in livestock systems. Developed under the auspices of the [FAO] Livestock, Environment and Development Initiative* (Davis: University of California, 2003).
36. David Pimentel and Marcia Pimentel, 'Sustainability of meat-based and plant-based diets and the environment', *American Journal of Clinical Nutrition* 2003: 78, pp. 660S–663S; D. Pimentel and M. Pimentel, *Food, Energy and Society* (New York: Taylor & Francis, 2008), pp. 68–70.
37. Jonathan Foley et al, 'Solutions for a Cultivated Planet', *Nature* 478 (2011), pp. 337–42; Weis, *The Ecological Hoofprint*, pp. 13–52. On obesity, see International Food Policy Research Institute, *Global Nutrition Report 2016* (Washington: IFPRI, 2016), p. 2; Franco Sassi, *Obesity and the Economics of Prevention* (Paris: OECD, 2010). On pets, research in the USA suggests that they consume about one-third as much animal-derived energy as humans do, i.e. about one quarter of the aggregate total, Gregory Okin, 'Environmental impacts of food consumption by dogs and cats', *PLOS One*, 2 August 2017. On waste, see Food and Agriculture Organisation of the UN (FAO), *Global Food Losses and Food Waste: extent, causes and prevention* (Rome: FAO, 2011); Tristram Stuart, *Waste: uncovering the global food scandal* (London: Penguin, 2009), especially pp. 91–3.
38. See Tony Weis, *The Global Food Economy: the battle for the future of farming* (London: Zed Books, 2007), especially pp. 163–72.
39. See Table 5-b. In 2011, agriculture, forestry and fishing was 190.7 million toe (2.1 per cent of the total); to this may be added a substantial proportion of the rows 'chemical and petrochemical', 'chemical feedstocks', and all forms of transport.
40. Estimates of agriculture's share of all greenhouse gas emissions are around 15–20 per cent, some higher; see FAO, *Livestock's Long Shadow* (Rome: FAO, 2006), p. xxi; WRI, *Navigating the Numbers*, pp. 72–3. Most of these are not from fossil fuel use, but from other sources including enteric fermentation (cows' farts, in non-specialist language) and manure spreading, which produce methane and nitrous oxide respectively. Deforestation, often driven by the search for agricultural land, also contributes to the greenhouse effect.
41. Allwood, Cullen et al, *Sustainable Materials*, p. 268. Design changes and re-use could reduce structural steel requirements to about two-fifths of current levels, this research found; improved quality rebar would yield further energy savings.
42. Chateau and Lapillonne, *Energy Demand*, p. 24; Anderer et al, *Energy in a Finite World*, vol. 2, p. 496; Johansson et al, *Global Energy*, pp. 653 and 658–9; Goldemberg et al, *Energy for a Sustainable World*, p. 121.
43. Johansson et al, *Global Energy*, pp. 657–9.
44. Anderer et al, *Energy in a finite world*, vol. 2, p. 495; J. van der Wal and K. Noorman, 'Analysis of Household Metabolic Flows' in Noorman and Uiterkamp, *Green Households?*, pp. 35–63, here pp. 37–9; Ruby Roy Dholakia, *Technology and Consumption: understanding consumer choices and behaviours* (New York: Springer, 2012), p. 47.
45. Goldblatt, *Sustainable Energy*, pp. 6–8.
46. Jennifer Clapp, 'The distancing', pp. 155–76; World Bank, *What a Waste: a global review of solid waste management* (Washington: World Bank, 2012), pp. 8–9.
47. WRAP, Refillable glass beverage container systems in the UK. RTL 0068 (Banbury: WRAP UK, June 2008); Nye, *Consuming*, p. 222.

48. Christina Reed, 'Dawn of the Plasticene', *New Scientist*, 31 January 2015, pp. 28–32; Marcus Eriksen et al, 'Plastic pollution in the world's oceans', *PLOS ONE* 9(12) (2014); J. Jambeck et al, 'Plastic waste inputs from land into the ocean', *Science*, 13 February 2015, pp. 768–71.

Chapter 5

1. The economic historian Angus Maddison estimated US GDP at 27.2 per cent of the world total in 1950 and 24.3 per cent in 1960; www.ggdc.net/maddison/oriindex.htm. See also Christopher Chase-Dunn et al, 'The Trajectory of the United States in the World-System: a quantitative reflection', *Sociological Perspectives* 48:2 (2005), pp. 233–54.
2. E. Robinson and G. Daniel, 'The World's Need for a New Source of Energy', in *Proceedings of the International Conference on the Peaceful Uses of Atomic Energy. Vol. 1. The World's Requirements for Energy: the role of nuclear energy* (New York: UN, 1956), pp. 38–49; O. F. Thompson, 'Oil as a Factor in Future Energy Supplies', *Transactions of the Fifth World Power Conference* (Vienna: Osterrechisches Nationalkomitee der Weltkraftkonferenz, 1957), vol. 3, pp. 463–84.
3. Chevron, Exxon's predecessor Standard Oil of New Jersey, Mobil, Texaco and Gulf Oil were US-based; BP and Royal Dutch Shell were not.
4. Debeir et al, *In the Servitude*, pp. 136–7; Goldemberg et al, *Energy for a sustainable world*, p. 5.
5. See e.g. Louis Turner, *Oil Companies and the International System* (London: George Allen & Unwin, 1983), pp. 38–67; Mitchell, *Carbon Democracy*, especially pp. 109–43; Robert Vitalis, *America's Kingdom: mythmaking on the Saudi oil frontier* (Stanford: Stanford University Press, 2007), especially pp. 1–26.
6. Gerard Brannon, *Energy Taxes and Subsidies* (Cambridge, Mass: Ballinger, 1974), pp. 25–45 and 172. Another researcher concluded that devices such as expensing for tax purposes of intangible drilling expenses, and percentage depletion, which allowed oil wells to be expensed many times over, served only 'political expediency'. Michael Yokell, 'The role of government in the development of solar energy', in Lewis Perelman, August Giebelhaus and Michael Yokell (eds.), *Energy Transitions: long-term perspectives* (Boulder, Co.: Westview Press, 1981), pp. 127–46). The first government-commissioned report on US energy production subsidies, published in 1978, estimated that since 1918 these had amounted to $217 billion in 1977 dollars, of which 54.3 per cent was for oil and gas. B. W. Cone et al, *An analysis of federal incentives used to stimulate energy production* (Richland: Pacific Northwest Laboratory, 1978).
7. Fiona Venn, *The Oil Crisis* (London: Longman, 2002), pp. 34–5; Turner, *Oil Companies*, pp. 43–9; Francisco Parra, *Oil Politics: a modern history of petroleum* (London: I.B. Tauris, 2010), pp. 17–20; Brannon, *Energy Taxes*, pp. 91–105.
8. These are proportions of fossil fuels consumed, *not* of total primary energy supply. Author's estimates from UN, *World Energy Supplies*, p. 42 and Darmstadter, *Energy in the World Economy*, p. 14. See also Joy Dunkerley, *Trends in Energy Use in Industrial Societies: an overview* (Washington: Resources for the Future, 1980), p. 131.
9. John Clark, *The Political Economy of World Energy: a twentieth-century perspective* (Hemel Hempstead: Harvester Wheatsheaf, 1990), pp. 104–7; Darmstadter, *Energy in the World Economy*, p. 18.
10. Rudolph and Ridley, *Power Struggle*, pp. 94–5; H. Wollmann et al., 'From public service to commodity'; Hausman et al, *Global Electrification*, pp. 234–9 and 248; Clark, *The Political Economy*, pp. 219–21; Patterson, *Transforming*, pp. 54–7.
11. Author's calculation, from data in Table 2 and www.worldeconomy.com.
12. Author's calculation, from UN, *World Energy Supplies*, p. 41, and Darmstadter, *Energy in the World Economy*, p. 34.

13. The research was commissioned by the US Electric Power Research Institute, and conducted with the cooperation of government agencies and industry bodies. Joel Darmstadter, Joy Dunkerley and Jack Alterman et al., *How Industrial Societies Use Energy* (Baltimore: Johns Hopkins University Press, 1977), especially pp. 185–207; J. Dunkerley, *Trends in Energy Use*, especially p. 57, p. 98 and pp. 130–32; Nathan Rosenberg, *Exploring the Black Box: technology, economics and history* (Cambridge: Cambridge University Press, 1994), p. 179.
14. When the IEA began publishing sectoral breakdowns in 1971, industry and construction accounted for 38 per cent of global final energy consumption, transport 23 per cent and residential 25 per cent. In the USA, industry and construction was 36 per cent, transport 31 per cent and residential 19 per cent. See Tables 5-b and 6-b.
15. Chateau and Lapillonne, *Energy Demand*, p.135 and p. 137.
16. Bruce Hannon, 'Energy, Labor and the Conserver Society', *Technology Review* 79:5, March/April 1977, pp. 1–12; Chateau and Lapillonne, *Energy Demand*, p. 146.
17. Chateau and Lapillonne, *Energy Demand*, p. 166 and p. 174; Andrew Wright, *The Motor Car Industry: a case study in structural change and its impact on energy demand* (Cambridge: Cambridge Energy Research Group, 1987), p. 11.
18. FAO, *The State of World Agriculture* 1951 (Rome: FAO, 1951), pp. 84–5 and 1968, pp. 44–6.
19. National Council of Applied Economic Research (NCAER), *Demand for Energy in India, 1960–1975* (New Delhi: Asia Publishing House, 1960), p. 94; Debeir et al., *In the Servitude*, pp. 147–8; Nye, *Consuming*, pp. 192–3. The capacity of motors in China was 49 million horsepower; human power was estimated at 30 million hp and animal power at 30 million hp. One horsepower is equal to 746 watts.
20. David Pimentel et al, 'Food Production and the Energy Crisis', *Science* no. 4111 (2 November 1973), pp. 443–9; Debeir et al., *In the Servitude*, pp. 144–6.
21. David Pimentel et al, 'Energy inputs in crop production in developing and developed countries', in Rattan Lal et al (eds.), *Food Security and Environmental Quality in the Developing World* (Boca Raton: CRC Press, 2003), pp. 129–51 and p. 145.
22. David Nye, *Consuming*, pp. 205–6; Tony Judt, *Postwar: a history of Europe since 1945* (London: Vintage, 2010), p. 339; NCAER, *Demand for Energy*, p. 50.
23. St Clair, *The Motorization*, pp. 160–64; Luger, *Corporate Power*, pp. 12–13; Goddard, *Getting There*, pp. 179–80 and 185–6; Huber, *Lifeblood*, pp. 40–41; McNeill and Painter, 'The global environmental footprint'.
24. F. Fisher, Z. Griliches and C. Kaysen, 'The Cost of Automobile Model Changes since 1949', *Journal of Political Economy* 70:5 (1962), pp. 433–51; Baran and Sweezy, *Monopoly Capital*, pp. 138–41 and 293–6; Berry and Fels, 'The Energy Cost of Automobiles'. See also Hughes, *The Bleeding Edge*, pp. 172–4.
25. Jane Jacobs, *The Death and Life of Great American Cities* (New York: Vintage Books, 1961), p. 7 and p. 351. See also Benoit Lefevre, 'Urban transport energy consumption'; Peter Newman, 'Reducing Automobile Dependence', *Environment and Urbanisation* 8:1 (1996), pp. 67–92.
26. Huber, *Lifeblood*, pp. 74–5 and p. 81; Goddard, *Getting There*, pp. 200–202.
27. Judt, *Postwar*, pp. 339–41; Goddard, *Getting There*, p. 226.
28. Chateau and Lapillonne, *Energy Demand*, pp. 20–24.
29. Judt, *Postwar*, pp. 385–7; Chateau and Lapillonne, *Energy Demand*, pp. 25–7.
30. Goldemberg et al, *Energy for a Sustainable World*, p. 111.
31. Chateau and Lapillonne, *Energy Demand*, pp. 35–6.
32. Cooper, *Air Conditioning America*, pp. 140–64; Raymond Arsenault, 'The End of the Long Hot Summer: the Air Conditioner and Southern Culture', *The Journal of Southern History*, vol. 50 no. 4 (1984), pp. 597–628, especially pp. 616–22.
33. Stanley Lebergott, *Pursuing Happiness: American consumers in the twentieth century* (Princeton: Princeton University Press, 1993), p. 113; Judt, *Postwar*, pp. 338–9;

F. Graham Pyett, *Priority Patterns and the Demand for Household Durable Goods* (Cambridge, Cambridge UP, 1964), pp. 40–41; van der Wal and Noorman, 'Analysis of Household Metabolic Flows', pp. 44–7.

34. Dholakia, *Technology and Consumption*, pp. 5–8; Ruby Roy Dholakia and Syagnik Banerjee, *Patterns of Durable Acquisition: empirical evidence from India* (Rhode Island: University of Rhode Island, College of Business Administration Working Paper series no. 11, 2011).
35. Rees, *Refrigeration Nation*, pp. 169–77; Johansson et al, *Global Energy*, p. 685; US Energy Information Administration (EIA), *25th Anniversary of the 1973 Oil Embargo: energy trends since the first major U.S. energy crisis* (Washington: EIA, 1998), Figure 26. For some indicators of refrigeration efficiency, see Rocky Mountain Institute, Home Energy Brief #8 Kitchen Appliances (2004), www.rmi.org.
36. Darmstadter, *Energy in the World Economy*, pp. 9–10.
37. Thane Gustafson, *Crisis Amid Plenty: the politics of Soviet energy under Brezhnev and Gorbachev* (Princeton: Princeton University Press, 1989), pp. 22–6; Dienes and Shabad, *The Soviet Energy System*, p. 34.
38. Dienes and Shabad, *The Soviet Energy System*, pp. 184–7 and p. 205.
39. Dienes and Shabad, *The Soviet Energy System*, pp. 19–21; M. P. Sacks, *Women's Work in the Soviet Union* (Praeger 1976), pp. 46–7.
40. Lewis Siegelbaum, 'Introduction', in Siegelbaum (ed.), *The Socialist Car: automobility in the Eastern Bloc* (Ithaca: Cornell University Press, 2011), pp. 1–13.
41. Goldemberg et al, *Energy for a sustainable world*, p. 12; Workshop on Alternative Energy Strategies, 'Energy and economic growth prospects for developing countries', in Smil and Knowland (eds.), *Energy in the developing world*, pp. 22–42, here 27–8.
42. Arjun Makhijani, 'Energy in the rural third world', in Smil and Knowland (eds.), *Energy in the developing world*, pp. 14–21; Cecelski et al, *Household energy*, pp. 7–8 and p. 73; World Bank, *Issues in Rural Electrification. Report no. 517* (Washington: World Bank, 1974), p. 23; Arjun Makhijani and Alan Poole, *Energy and Agriculture in the Third World* (Cambridge, Mass.: Bellinger, 1975), p. 22.
43. K. Openshaw, 'Woodfuel – a time for reassessment', in Smil and Knowland (eds.), *Energy in the developing world*, pp. 72–86, here p. 72; Cecelski et al, *Household energy*, pp. 15–21; Amulya Reddy and B. Sudhakar Reddy, 'Energy in a stratified society – case study of firewood in Bangalore', *Economic and Political Weekly*, 8 October 1983, pp. 1757–70.
44. Pachauri, *An Energy Analysis*, pp. 14–15; NCAER, *Demand for Energy*, p. 21, pp. 32–3 and p. 46.
45. Makhijani and Poole, *Energy and Agriculture*, pp. 20–24; Nick Cullather, *The Hungry World: America's Cold War Battle Against Poverty in Asia* (London: Harvard University Press, 2010), especially pp. 108–204.
46. V. Smil, *China's Past, China's Future: energy, food, environment* (London: Routledge, 2004), pp. 10–11; Sanchez, *The Hidden Energy Crisis*, p. 62; Pan Jiahua et al, *Rural Electrification in China 1950–2004. Working Paper 60* (Stanford: Stanford University Programme on Energy and Sustainable Development, 2006), p. 26.
47. Ruth Schwartz Cowan, *More Work for Mother: the ironies of household technology from the open hearth to the microwave* (New York: Basic Books, 1983), p. 201 and pp. 211–13; Ursula Huws, *Labor in the Global Digital Economy* (New York: Monthly Review Press, 2014), pp. 7–8.
48. Joann Vanek, 'Time spent in housework', *Scientific American*, 1 November 1974, pp. 116–20; Valerie A. Ramey, 'Time spent in home production in the twentieth-century United States: new estimates from old data', *Journal of Economic History* 69:1 (March 2009), pp. 1–47. The figures quoted are estimates of average weekly hours spent in home production by women aged 18–64, from table 6a, p. 27. See also Heidi

Hartmann, 'The family as the locus of gender, class and political struggle: the example of housework', *Signs* 6:3 (Spring 1981), pp. 366-94.
49. Donald Filtzer, *Soviet workers and de-Stalinization: the consolidation of the modern system of Soviet production relations* (Cambridge: Cambridge University Press, 1992), pp. 196-204.
50. Harold Wilhite, 'A Socio-Cultural Analysis of Changing Household Electricity Consumption in India' in D. Spreng et al (eds.), *Tackling Long Term Global Energy Problems: the contribution of social science* (Dordrecht: Springer 2012), pp. 97-113.
51. The argument is made in: Jeremy Greenwood, Ananth Seshadri and Mehmet Yorukoglu, 'Engines of Liberation', *The Review of Economic Studies* 72:1 (2005), pp. 109-33.
52. Cowan, *More Work for Mother*, p. 183; Lebergott, *Pursuing Happiness*, p. 60.

Chapter 6

1. Author's calculations from BP statistics.
2. Prices fell from $2.08/barrel in 1959 to $1.80/barrel in 1970 (BP *Statistical Review*), but the fall was steeper in real terms (i.e. adjusted for inflation). Prices fell by half, according to Peter Odell's estimates. Odell, *Oil and World Power* (London: Penguin, 1986), p. 142.
3. US oil imports rose from 67.7 mt in 1970 to 165.8 mt in 1973 and hit a peak of 333.3 mt in 1979. They fell back to 178.3 mt in 1982, and rose gradually thereafter, regaining their 1979 level only in 2001 (EIA data).
4. Odell, *Oil and World Power*, pp. 114-44 and p. 220; Parra, *Oil Politics*, pp. 68-88; Turner, *Oil Companies*, pp. 52-3; Robert Lieber, *The Oil Decade: conflict and cooperation in the west* (New York: Praeger, 1983), pp. 44-5.
5. Parra, *Oil Politics*, pp. 146-85; Ian Skeet, *Opec: twenty-five years of prices and politics* (Cambridge: Cambridge University Press, 1988), pp. 58-96; Odell, *Oil and World Power*, pp. 217-31; Turner, *Oil Companies*, pp. 125-48; Venn, *The Oil Crisis*, pp. 7-9 and 17-21.
6. Philip Armstrong, Andrew Glyn and John Harrison, *Capitalism Since World War II: the making and breaking of the long boom* (Oxford: Basil Blackwell, 1991), pp. 169, 210-15 and 225-9; Clark, *The Political Economy*, pp. 236-9.
7. Parra, *Oil Politics*, pp. 215-39; Skeet, *Opec*, pp. 157-77 and 209-11; Bruce Scott, 'OPEC, the American scapegoat', *Harvard Business Review*, January-February 1981.
8. Rajendra Pachauri, *The Political Economy of Global Energy* (Baltimore: Johns Hopkins University Press, 1985), pp. 16-19.
9. Armstrong et al, *Capitalism Since 1945*, pp. 224-5; Clark, *The Political Economy*, p. 259.
10. IEA, *World Energy Outlook 1982* (Paris: IEA, 1982), pp. 82-3. See also Clare Doblin, *The Growth of Energy Consumption and Prices in the USA, FRG, France and the UK, 1950-1980* (Laxenburg: IIASA, 1982).
11. Clark, *The Political Economy*, pp. 264-5; IEA, *World Energy Outlook 1982*, pp. 81-2.
12. IEA *World Energy Outlook 1982*, pp. 97-8; Chateau and Lapillonne, *Energy demand*, p. 108; David Nye, *Consuming*, pp. 220-21; IEA, *Thirty Years*, p. 17.
13. IEA, *World Energy Outlook 1982*, pp. 91-2; Goldemberg et al, *Energy for a sustainable world*, pp. 101-2 and 142; Merton Peck and John Beggs, 'Energy Conservation in American Industry', in John Sawhill and Richard Cotton (eds.), *Energy Conservation: successes and failures* (Washington: The Brookings Institution, 1986), pp. 59-93, here pp. 60-61.
14. Udo Simonis, 'Industrial restructuring in industrial countries', in Robert Ayres and Udo Simonis (eds.), *Industrial metabolism: restructuring for sustainable development* (Tokyo: UN University Press, 1994), pp. 31-54, here pp. 47-9.

15. Goldemberg et al, *Energy for a sustainable world*, pp. 99–103; IEA, *Thirty Years*, p. 14; Clark, *The Political Economy*, pp. 258–60.
16. IEA, *World Energy Outlook 1977* (Paris: IEA, 1977), p. 10; Lieber, *The Oil Decade*, p. 3 and pp. 54–5; Parra, *Oil Politics*, pp. 198–204.
17. Patterson, *Transforming Electricity*, pp. 66–7.
18. John Clark, *The Political Economy of World Energy*, pp. 259–60.
19. Meg Jacobs, *Panic at the Pump: the energy crisis and the transformation of American politics in the 1970s* (New York: Hill and Wang, 2016), pp. 36–48 and 135–8.
20. Nemet, 'Automobile fuel', pp. 180–82; Margo Oge, *Driving the Future: combating climate change with cleaner, smarter cars* (New York: Arcade, 2015), p. 103; Luger, *Corporate Power*, pp. 94–5; Wright, *The Motor Car Industry*, pp. 19–21.
21. Jacobs, *Panic at the Pump*, pp. 28–36; Desta Mebratu, 'Sustainability and sustainable development: historical and conceptual review', *Environmental Impact Assessment Review* 1998 (18), pp. 493–520.
22. Lovins gave testimony to a joint hearing of the Select Committee on Small Business and the Committee on Interior and Insular Affairs, 9 December 1976. See Lovins, *The Energy Controversy*, pp. 15–34; Lovins et al, *Small is Profitable*, p. 38.
23. Jacobs, *Panic at the Pump*, pp. 162–91; Rosenberg, *Exploring the Black Box*, pp. 181–2; Nye, *Consuming*, pp. 219–20 and 234–45; Patterson, *Transforming Electricity*, pp. 72–3.
24. Jose Goldemberg et al, 'Basic Needs and Much More with One Kilowatt per Capita', *Ambio* 14, no. 4/5 (1985), pp. 190–200.
25. Nicolas Jequier (ed.), *Appropriate technology: problems and promises* (Oxford: Oxford University Press, 1976); Nicolas Jequier and Gerard Blanc, *The World of Appropriate Technology: a quantitative analysis* (Paris: OECD, 1983), pp. 40–42 and 165–7; Malcolm Hollick, 'The Appropriate Technology movement and its literature: a retrospective', *Technology in Society* 4 (1982), pp. 213–29.
26. *Energy Policy*, June 1974, especially P. Chapman, 'Energy costs: a review of methods', pp. 91–103; see also David Murphy, 'The implications of the declining energy return'. William Nordhaus, 'The demand for energy: an international perspective', in W. Nordhaus (ed.), *Proceedings of the Workshop on Energy Demand* (Laxenburg: IIASA, 1976), pp. 511–87, here p. 511.
27. *World Energy Demand: the full reports to the Conservation Commission of the World Energy Conference* (New York: IPC/WEC, 1978), especially pp. 9–16; Jean-Romain Frisch (ed.), *Energy 2000-2020: world prospects and regional stresses (World Energy Conference Conservation Commission)* (London: Graham & Trotman, 1983), in which there is 'no analysis of demand by sectors or of useful energy' (p. xxiii).
28. Workshop on Alternative Energy Strategies (WAES), *Energy: Global Prospects 1985-2000* (Boston: MIT, 1977); Paul Basile (ed.), *Energy Supply-Demand Integrations to the Year 2020* (London: MIT Press, 1977); Anderer et al, *Energy in a finite world: paths to a sustainable future* (Laxenburg: IIASA, 1981); Goldemberg et al, *Energy for a sustainable world*, especially pp. 15–17.
29. Patterson, *Transforming Electricity*, p. 67; Vaclav Smil and William Knowland, 'Energy in the Developing World', in Smil and Knowland (eds.), *Energy in the developing world*, pp. 5–13, here note 15 on p. 12.
30. J. Goldemberg, 'Brazil: Energy Options and Current Outlook', *Science*, 14 April 1978, pp. 158–64; J. Goldemberg, 'Energy Issues and Policies in Brazil', *Annual Review of Energy* 7 (1982), pp. 139–74; Clark, *The Political Economy* , p. 281; Goldemberg et al, *Energy for a Sustainable World*, pp. 240–54.
31. Paul Hallwood and Stuart Sinclair, *Oil, Debt and Development: OPEC in the Third World* (London: George Allen and Unwin, 1981), pp. 56–7, 74–81 and 84–9; Akin Iwayemi, 'Energy development and sub-Saharan African economies in a global perspective', *OPEC Review* XVII:1 (Spring 1993), pp. 29–45.
32. Clark, *The Political Economy* , p. 283; Hallwood and Sinclair, pp. 68–9.

33. Oil Change International, *Low Hanging Fruit: fossil fuel subsidies, climate finance and sustainable development* (Washington: Oil Change International, 2012), p. 32; David Victor, *Untold Billions: fossil fuel subsidies, their impacts and the paths to reform* (Geneva: GSI, 2010), p. 19.
34. Amulya Reddy and B. Sudhakar Reddy, 'Energy in a stratified society – case study of firewood in Bangalore', *Economic and Political Weekly*, 8 October 1983, pp. 1757–70.
35. Clark, *The Political Economy*, pp. 281–2; P. D. Henderson, 'Energy resources, consumption and supply in India', in Smil and Knowland (eds.), *Energy in the developing world*, pp. 172–93; Roger Revelle, 'Energy use', pp. 193–210; Cecelski et al, *Household energy*, p. 14.
36. Erik Eckholm, *The Other Energy Crisis: Firewood* (Washington: Worldwatch Institute, 1975); E. Eckholm et al., *Fuelwood: the energy crisis that won't go away* (Washington: IIED, 1982), p. 5; Michael Arnold et al, *Fuelwood Revisited: what has changed in the last decade?* (Bogor, Indonesia: CIFOR, 2003), pp. 3–6.
37. V. Smil, 'China's energetics: a systems analysis', in Smil and Knowland (eds.), *Energy in the developing world*, pp. 113–44, here p. 123; Smil, *China's Past*, pp. 39–50; Sanchez, *The Hidden Energy Crisis*, p. 62.
38. Clark, *The Political Economy*, p. 294; Yergin, *The Quest*, p. 135.

Chapter 7

1. J. H. Williams and R. Ghanadan, 'Electricity reform in developing and transition countries: a reappraisal', *Energy* (31) 2006, pp. 815–44; Clark, *The Political Economy*, p. 116 and pp. 219–21; Anne Salda, *Historical Dictionary of the World Bank* (London: Scarecrow, 1997), pp. 71–4.
2. Patterson, *Transforming Electricity*, pp. 57–8.
3. World Bank, *Rural Electrification* (Washington: World Bank, 1975), p. 3, Cecelski et al, *Household Energy*, p. 14; Bose, *Money, Energy and Welfare*, p. 99.
4. Hugh Collier, *Developing Electric Power: thirty years of World Bank Experience* (Washington: World Bank, 1984), pp. 124–5.
5. Cecelski et al, *Household Energy*, pp. 14–15 and 71–4; World Bank, *Rural Electrification*, p. 17.
6. S. Karekezi and A. Sihag/Global Network on Energy for Sustainable Development (GNESD) 'Energy Access' Working Group, *Final Synthesis/Compilation Report* (Fredriksborgvej: GNESD, 2004), especially pp. 32–3. Other reports that support these conclusions include chapters in Navroz Dubash (ed.), *Power politics: equity and environment in electricity reform* (Washington: World Resources Institute, 2002), and Neiji Wamukonya (ed.), *Electricity Reform: social and environmental challenges* (Nairobi: UNEP, 2003).
7. Makhijani and Poole, *Energy and Agriculture*, p. 22; Goldemberg et al, *Energy for a Sustainable World*, p. 197; Cecelski et al, *Household Energy*, pp. 26–7 and 73–5.
8. World Bank, *The Welfare Impact of Rural Electrification: a reassessment of the costs and benefits* (Washington: World Bank, 2008), p. xv; Sanchez, *The Hidden Energy Crisis*, pp. 13–15.
9. Santhakumar, *Analysing Social Opposition*, p. 125; Adinife Azodo, 'Electric power supply, main source and backing: a survey of residential utilization features', *International Journal of Research Studies in Management* 3:2 (2014), pp. 87–102.
10. Jonathan Coopersmith, *The Electrification of Russia 1880–1926* (Ithaca: Cornell University Press, 1992), pp. 45–6; V. Gvozdetskii, 'Plan GOELRO. Mify i real'nost', *Nauka i zhizn'* no. 5, 2001. The managers of Goelro included Leonid Krasin, Gleb Krzhizhanovskii and Petr Smidovich, all senior members of the Communist party.

11. *Trudy 8-oi Vserossiiskogo elektrotekhnicheskogo s"ezda v Moskve 1–10 oktiabria 1921 goda. Vypusk 1. Elektrifikatsiia Rossii* (Moscow: Partizdat, 1921), especially pp. 72–90; Coopersmith, *The Electrification*, pp. 183–91.
12. Coopersmith, *The Electrification*, pp. 258–9. It is to the credit of the Russian electricity holding company United Energy Systems that it funded research by historians on the role of prison labour in power station construction. See: Oleg Khlevniuk et al, *Zakliuchennyie na stroikakh kommunizma: Gulag i ob"ekty energetiki v SSSR* (Moscow, Rosspen, 2008). On investment, p. 18.
13. Nebeker, *Dawn*, p. 109; Darmstadter, *Energy in the World Economy*, pp. 9–10.
14. Dienes and Shabad, *The Soviet Energy System*, pp. 184–7; Nove, *An Economic History*, p. 234; Nebeker, *Dawn*, pp. 116–17; Abel Aganbegyan, *The Challenge: Economics of Perestroika* (London: Hutchinson, 1988), p. 51.
15. Dienes and Shabad, *The Soviet Energy System*, p. 187 and p. 207; David Wilson, *The demand for energy in the Soviet Union* (London, Crooms Helm, 1983), pp. 45–7.
16. Table 3 (p. 42); UNDESA Population Division statistics, www.un.org/en/development/desa/population; Chi Zhang and Thomas Heller, 'Reform of the Chinese electric power market', in David Victor and Thomas Heller (eds.), *The Political Economy of Power Sector Reform* (Cambridge: Cambridge University Press, 2007), pp. 76–108. It is frequently stated that 900 million people have been brought on to the electricity network since 1949. Many of these people were added by population growth: China's population rose by 745 million between 1949 and 2000, and a further 108 million between 2000 and 2015.
17. Table 6-b (p. 63); Subhes Bhattacharyya and Sanusi Ohiare, 'The Chinese electricity access model for rural electrification', *Energy Policy* 49 (2012), pp. 676–87.
18. Table 6-a (p. 62). Coal and peat, as a proportion of the inputs to transformation, from which electricity and heat are produced, were 84 per cent in 1971, 81 per cent in 1991 and 87 per cent in 2011.
19. William Clarke, 'China's electric power industry', in Smil and Knowland (eds.), *Energy in the Developing World*, pp. 145–66, here p. 147; Zhaohong Bie and Yanling Lin, 'An Overview of Rural Electrification in China: history, technology and emerging trends', *IEEE Electrification* magazine, March 2015, pp. 36–47.
20. Pan Jiahua et al, *Rural Electrification in China 1950–2004* (Stanford: Stanford University Programme on Energy and Sustainable Development, Working Paper 60, 2006); Wuyuan Peng and Jiahua Pan, 'Rural Electrification in China: History and Institution', *China and World Economy* 14:1 (2006), pp. 71–84.
21. Clarke, 'China's electric power'; Bhattacharyya and Ohiare, 'The Chinese electricity access model'; Zhang and Heller, 'Reform'; Xu Yi-chong, *Electricity Reform in China, India and Russia: the World Bank template and the politics of power* (Cheltenham: Edward Elgar, 2004), especially pp. 107–11 and 186–202.
22. Peng and Pan, 'Rural Electrification'; Bie and Lin, 'An Overview'.
23. Sunila Kale, *Electrifying India*, pp. 36–7.
24. In 2011 the level of household electricity access was put at 67 per cent by the census authorities, 71 per cent by the Central Electricity Authority and 74 per cent by the national surveys. Shonali Pachauri, 'Household electricity access'.
25. NCAER, *Demand for Energy*, p. 46; Makhijani and Poole, *Energy and Agriculture*, pp. 20–25; Navroz Dubash and Sudhir Rajan, 'India: Electricity Reform Under Political Constraints', in N. Dubash (ed.), *Power Politics*, pp. 51–73; Sudhir Katiyar, 'Political economy of electricity theft in rural areas: a case study from Rajasthan', *Economic and Political Weekly*, 12 February 2005; Roger Revelle, 'Energy use'.
26. Pachauri, *An Energy Analysis*, pp. 34–6; J.K. Mathur and D. Mathur, 'Dark Homes and Smoky Hearths: rural electrification and women', *Economic and Political Weekly*, 12 February 2005; N. Sreekumar and S. Dixit, 'Challenges in Rural Electrification', *Economic and Political Weekly*, 22 October 2011.

27. Sunila Kale, *Electrifying India*, pp. 62–99; Williams and Ghanadan, 'Electricity reform'; Dubash and Rajan, 'India: Electricity reform'; Kalpana Dixit, *Political Economy of Electricity Distribution in Maharashtra* (Delhi: Centre for Policy Research and Regulatory Assistance Project, 2017), p. 11 and p. 18.
28. Pachauri, *An Energy Analysis*, pp. 52–3; Navroz Dubash, *Tubewell Capitalism: groundwater development and agrarian change in Gujarat* (Oxford: Oxford University Press, 2002), especially pp. 250–51; Siddharth Sareen, *What Powers Success on the Ground? The Gradual Reform of Electricity Distribution in Gujarat* (Delhi: Centre for Policy Research and Regulatory Assistance Project, 2017).
29. Kale, *Electrifying India*, pp. 100–135; Dubash and Rajan, 'India: Electricity reform'; A. Sihag, Neha Misra and Vivek Sharma, 'Impact of power sector reform on the poor: case studies of South and South-East Asia', *Energy for Sustainable Development* 8:4 (2004), pp. 54–67, here p. 61.
30. Kale, *Electrifying India*, pp. 136–59.
31. Ben Arimah, 'Electricity consumption in Nigeria: a spatial analysis', *OPEC Review* XVII:1 (1993), pp. 63–82, here p. 64; GOPA International Energy Consultants, *The Nigerian Energy Sector* (Berlin: GIZ, November 2014), p. 35; A.S. Aliyu et al, 'Nigeria electricity crisis', *Energy* 61 (2013), pp. 354–76; Akin Iwayemi, 'Nigeria's dual energy problems: policy issues and challenges', *IAEE Newsletter*, 4th quarter 2008, pp. 17–21.
32. GOPA, *The Nigerian Energy Sector*, p. 38, 45 and 130; Aliyu et al, 'Nigeria electricity crisis', pp. 354 and 357.
33. A. Olukoju, *Infrastructure Development and Urban Facilities in Lagos, 1861–2000* (Open Edition Books, 2003), p. 86.
34. Arimah, 'Electricity consumption'; Olukoju, *Infrastructure Development*, p. 32; Ignatius Ukpong, 'Social and economic infrastructure', in F. Akin Olaloku (ed.), *Structure of the Nigerian Economy* (New York: St Martin's Press, 1979), pp. 68–99.
35. Olokuju, *Infrastructure Development*, pp. 32–4.
36. A. Olokuju, '"Never Expect Power Always": electricity consumers' response to monopoly, corruption and inefficient services in Nigeria', *African Affairs* 103 (2004), pp. 51–71; David Hall, *Electrifying Africa: power through the public sector* (London: PSIRU, 2007), pp. 6–7; Akin Iwayemi, 'Nigeria's dual energy problems'.
37. Oil deposits often yield natural gas that comes to the surface together with the oil (associated gas). Long-standing oil industry practice, dating from before the widespread use of natural gas in power generation, was to flare the gas (burn it at the well-head), or, worse still from a greenhouse impact point of view, vent it. Methane's greenhouse effect is many times greater than carbon dioxide's: 86 times greater over 20 years, or 34 times greater over 100 years. (See 'How Bad of a Greenhouse Gas is Methane?', *Scientific American*, 22 December 2015, for a summary.) So pressure has mounted on oil companies to invest in infrastructure to capture and transport the gas.
38. Global Gas Flaring Reduction Partnership web site, www.worldbank.org/en/programs/gasflaringreductions, S. Pirani, 'Gas flaring reduction progress reviewed', *Gas Matters*, December 2012; Olokuju, '"Never Expect Power Always"'; Shell Nigeria, *Gas Flaring* (information sheet, 2014). On conditions in the oil producing region, see many reports by NGOs, e.g. Platform London, *Counting the cost: corporations and human rights abuses in the Niger Delta* (London: Platform London, 2011); Amnesty International, *Petroleum, Pollution and Poverty in the Niger Delta* (London: Amnesty International, 2009).
39. David McDonald, 'Electric capitalism', in D. McDonald (ed.), *Electric Capitalism: recolonising Africa on the power grid* (London: Earthscan, 2009), pp. 1–49.
40. Renfrew Christie, *Electricity, Industry and Class in South Africa* (New York: New York State University Press, 1984), pp. 5–103; Leonard Gentle, 'Escom to Eskom', in D. McDonald (ed.), *Electric Capitalism*, pp. 50–72, here pp. 52–6; Wendy Annecke,

'Still in the shadows', in D. McDonald (ed.), *Electric Capitalism*, pp. 288-320, here p. 302.

41. Ben Fine and Zavareh Rustomjee, *The Political Economy of South Africa: from minerals-energy complex to industrialisation* (London: C. Hurst and co., 1996), pp. 148-58; Gentle, 'Escom to Eskom'; Peter van Heusden, 'Discipline and the new "logic of delivery"' in McDonald (ed.), *Electric Capitalism*, pp. 229-47; Wendy Annecke, 'Still in the Shadows'.
42. McDonald, 'Electric Capitalism'; Stephen Greenberg, 'Market liberalisation and continental expansion', in McDonald (ed.), *Electric Capitalism*, pp 73-108; Julia Philpott and Alix Clark, 'South Africa: Electricity Reform with a Human Face?', in Dubash (ed.), *Power Politics*, pp. 139-55; Patrick Bond and Trevor Ngwane, 'Community Resistance to Energy Privatisation in South Africa' in Kolya Abramsky (ed.), *Sparking a Worldwide Energy Revolution: social struggles in the transition to a post-petrol world* (Oakland: AK Press, 2010), pp. 197-207.
43. McDonald, 'Electric Capitalism'.

Chapter 8

1. Bjorn Larsen and Anwar Shah, *World Fossil Fuel Subsidies and Global Carbon Emissions*, Working Paper 1002 (Washington: World Bank, 1992); World Resources Institute, *World Resources 1990-91* (New York: Oxford University Press, 1990), table 24.1 on pp. 346-7; *BP Statistical Review*.
2. Armstrong et al, *Capitalism Since 1945*, pp. 234-8.
3. David McNally, 'From financial crisis to world slump: accumulation, financialisation and the global slowdown', *Historical Materialism* 17 (2009), pp. 35-83; Angus Maddison, *The World Economy: a millenial perspective* (Paris: OECD, 2006), p. 136; Michael Roberts, *The Long Depression* (Chicago: Haymarket, 2016), pp. 59-64.
4. McNally, 'From financial crisis', p. 60; Maddison, *The World Economy*, p. 136.
5. Robert Mabro, 'Introduction', in R. Mabro (ed.), *The 1986 Oil Price Crisis: economic effects and policy responses* (Oxford: Oxford University Press, 1988), pp. 1-14, here pp. 13-14; John Sawhill and Richard Cotton, 'Introduction' in Sawhill and Cotton, *Energy Conservation*, pp. 1-17; Paul Stevens, 'Oil markets and the future', in Dieter Helm (ed.), *The New Energy Paradigm* (Oxford: Oxford University Press, 2007), p. 128; Clark, *The Political Economy*, pp. 326-7; Debeir et al, *In the Servitude*, pp. 139-40; Berck and Roberts, 'History of Prices'.
6. Patterson, *Transforming Electricity*, p. 74; Lovins, *Small is Profitable*, p. 31; Per Högselius, *Red Gas: Russia and the origins of European energy dependence* (New York: Palgrave Macmillan, 2013), pp. 184-202.
7. See reports by the International Institute for Applied Systems Analysis in 1981 (Anderer et al, *Energy in a finite world*, vol. 1, especially pp. 43-62 and 196-7), and the World Energy Conference conservation commission in 1983 (J.-R. Frisch (ed.), *Energy 2000-2020: world prospects and regional stresses* (London: Graham & Trotman, 1983), pp. xxx-xxxi and pp. 108-13). See also Goldemberg et al, *Energy for a Sustainable World*, pp. 15-17.
8. William Hogan, 'Patterns of Energy Use' in Sawhill and Cotton (eds.), *Energy Conservation*, pp. 19-58, here p. 19.
9. James Jensen, 'Oil and Energy Demand: outlook and issues', in Mabro (ed.), *The 1986 Oil Price Crisis*, pp. 17-40, here pp. 22-3.
10. Netherlands Environmental Assessment Agency, *Trends in Global CO_2 Emissions: 2013 Report* (Amsterdam: NEAA, 2013), p. 35; UNDP, *World Energy Assessment*, p. 35 and p. 57; Berck and Roberts, 'History of prices'; IEA, *Thirty Years*, pp. 42, 72 and 78-9; David Nye, *Consuming*, pp. 237-8.

11. World Resources Institute, *The Weight of Nations: material outflows from industrial economies* (Washington: WRI, 2000), pp. vi–vii; UNDP, *World Energy Assessment*, p. 178.
12. Jensen, 'Oil and Energy Demand', pp. 27–32.
13. Rosenberg, *Exploring*, p. 186; Wright, *The Motor Car Industry*, pp. 5–9 and 40–42; Allwood et al., *Sustainable Materials*, pp. 30 and 32.
14. Spreng, *Net-Energy Analysis*, p. 62; UNDP, *World Energy Assessment*, p. 186 and p. 57.
15. Francois-Xavier Devetter and Sandrine Rousseau, 'Working Hours and Sustainable Development', *Review of Social Economy* 69:3 (2011), pp. 333–55; Allwood et al, 'Industry 1.61803'.
16. Juliet Schor, 'Prices and quantities: unsustainable consumption and the global economy', *Ecological Economics* 55(2005), pp. 309–20. Schor counted children between 0 and 12 to produce this 'conservative' estimate. According to toy manufacturers, although some toys are consumed by adults, children between 10 and 12 consume very few.
17. IEA, *Thirty Years*, pp. 15–16, 83–4 and 102; Fouquet, *Heat, Power and Light*, p. 87.
18. Goldblatt, *Sustainable Energy Consumption*, pp. 6–8.
19. IEA, *Thirty Years*, p. 17.
20. EIA, *25th Anniversary*, figure 4; A. Downs, *Stuck in Traffic: coping with peak-hour traffic congestion* (Washington: Brookings Institution, 1992), p. 68; Richard Homan, 'Gas peddled at a wide range of prices', *The Washington Post*, 12 January 1991, p. A7.
21. An et al, *Global Overview*, p. 5; Frank von Hippel, *Citizen Scientist* (New York: American Institute of Physics, 1991), p. 174; F. von Hippel and A. Gray, 'The fuel economy of light vehicles', *Scientific American*, May 1981, pp. 48–59.
22. Oge, *Driving*, pp. 103–7; Osamu Kimura, 'The role of standards', in Grubler and Wilson (eds.), *Energy Technology Innovation*, pp. 231–43.
23. Downs, *Still Stuck in Traffic*, pp. 43–9 and p. 102; 'City of cars', *Economist*, 29 April 2017.
24. Amanda Root, 'Transport and communication', in A. H. Halsey and Josephine Webb (eds.), *Twentieth Century British Social Trends* (Basingstoke: Macmillan, 2000), pp. 437–68.
25. UNDP, *World Energy Assessment*, pp. 180–81.
26. McNeill and Engelke, *The Great Acceleration*, p. 23; Olukoju, *Infrastructure Development*, p. 70; Marcia Lowe, 'Rethinking Urban Transport', in Lester Brown et al, *State of the World 1991* (New York: W.W. Norton, 1991), pp. 56–73, here p. 71 (quoted); L. Siegelbaum, 'Introduction', in Siegelbaum (ed.), *The Socialist Car*; IEA, *Worldwide Trends*, p. 63.
27. Ross Garnaut, Frank Jotzo and Stephen Howes, 'China's rapid emissions growth and global climate change policy', in Ligang Song and Wing Thye Woo (eds.), *China's Dilemma: economic growth, the environment and climate change* (Washington: Brookings Institution Press, 2008), pp. 170–89; Vaclav Smil, *China's Past*, pp. 23–4.
28. Rajendra Pachauri, Mala Damodaran and Himraj Dang, 'Industrial restructuring in India', in Robert Ayres and Udo Simonis (eds.), *Industrial Metabolism*, pp. 55–77, here pp. 63–72; Pachauri, *An Energy Analysis*, pp. 52–4.
29. Jayant Sathaye, Andre Ghirardi and Lee Schipper, 'Energy Demand in Developing Countries: a sectoral analysis of recent trends', *Annual Review of Energy* 12 (1987), pp. 253–81; Ashwini Kulkarni, Girish Sant and J.G. Krishnayya, 'Urbanisation in Search of Energy in three Indian cities', *Energy* 19:5 (1994), pp. 549–60. See also: Douglas Barnes et al, *The Urban Household Energy transition: social and environmental impacts in the developing world* (Washington: Resources for the Future, 2005).
30. Harold Wilhite, 'A Socio-Cultural Analysis'; H. Wilhite, *Consumption and the transformation of everyday life: a view from South India* (Basingstoke: Palgrave Macmillan, 2008), pp. 121–5.

31. Pachauri, *An Energy Analysis*, pp. 164-7; Angus Deaton and Jean Dreze, 'Poverty and inequality in India: a re-examination', *Economic and Political Weekly*, 7 September 2002, pp. 3729-48.
32. Beverley Hooper, '"Flower vase and housewife": women and consumerism in post-Mao China' in Krishna Sen and Maila Stivens (eds.), *Gender and Power in Affluent Asia* (London: Routledge, 1998), pp. 167-93.
33. Pan Jiahua et al, *Rural Electrification*; Kirk Smith et al, 'One Hundred Million Improved Cookstoves in China: how was it done?', *World Development* 21:6 (1993), pp. 941-61.
34. The Sahel includes parts of northern Senegal, southern Mauritania, central Mali, northern Burkina Faso, Niger; the extreme north of Nigeria, Cameroon and the Central African Republic; central Chad, central and southern Sudan, the extreme north of South Sudan, Eritrea, and the extreme north of Ethiopia.
35. Akin Iwayemi, 'Energy development', pp. 31-2 and 35-7; Table 3, p. 42.
36. Huber, *Lifeblood*, p. 118; Nye, *Consuming*, pp. 235-6; Jacobs, *Panic at the Pump*, pp. 271-81; Lieber, *The Oil Decade*, pp. 9-10; Rudolph and Ridley, *Power Struggle*, p. 11 and pp. 16-17; Luger, *Corporate Power*, pp. 127-32; Peter Jacques, Riley Dunlap and Mark Freeman, 'The Organization of Denial: Conservative think tanks and environmental scepticism', *Environmental Politics* 17:3 (2008), pp. 349-85.
37. David Harvey, *A Brief History of Neoliberalism* (Oxford: Oxford University Press, 2005), pp. 12-19; Brendan Martin, *In the Public Interest? Privatisation and Public Sector Reform* (London: Zed Books, 1993), pp. 47-52.
38. Daniel Pope, *Nuclear Implosions: the rise and fall of the Washington public power supply system* (Cambridge: Cambridge University Press, 2008); Rudolph and Ridley, *Power Struggle*, pp. 8-9.
39. Dieter Plehwe/ICEF, *Energy Privatization and 'Deregulation'* (Brussels: ICEF, 1992), pp. 23-4.
40. Doug Koplow, 'Subsidies to Energy Industries', in Cutler Cleveland (ed.), *Encyclopaedia of Energy*. Vol 5 (Amsterdam: Elsevier, 2004).
41. Brannon, *Energy Taxes*, especially pp. 11-22; Cone et al, *An analysis of federal incentives used to stimulate energy production*; R.J. Cole et al, *An analysis of federal incentive used to stimulate energy consumption* (Richland: Pacific Northwest Laboratory, 1981), pp. 35-45; P.C. Spewak, 'Analyzing the effect of economic policy on solar markets', in R.E. West and Frank Kreith (eds.), *Economic Analysis of Solar Thermal Energy Systems* (London: MIT Press, 1988), pp. 167-204, p. 174.
42. H. Richard Heede, Richard Morgan and Scott Ridley, *The Hidden Costs of Energy: how taxpayers subsidise energy development* (Washington: Centre for Renewable Resources, 1985), pp. 2-6.
43. Mark Kosmo, *Money to Burn* (Washington: WRI, 1987).
44. Myers, *Perverse subsidies*, p. xvii. I have not found any research devoted specifically to the global level of road transport subsidies.
45. Bonneuil and Fressoz, *The Shock*, pp. 72-9.
46. Alan Hecht and Dennis Tirpak, 'Framework agreement on climate change: a scientific and policy history', *Climatic Change* 29 (1995), pp. 371-402 here pp. 371-2; Spencer Weart, *The Discovery of Global Warming* (London: Harvard University Press, 2003), pp. 117-24 and 145; Daniel Bodansky, 'Prologue to the Climate Change Convention', in Irving Mintzer and J. Leonard (eds,), *Negotiating Climate Change: the inside story of the Rio convention* (Cambridge: Cambridge University Press 1994), pp. 45-74, here pp. 46-7.
47. Weart, *The Discovery*, pp. 125-7 and 133-7; Valerie Masson-Delmotte, 'Ice with everything', in David Walton (ed.), *Antarctica: global science from a frozen continent* (Cambridge: Cambridge University Press, 2013), pp. 67-101, here p. 87 and Hans Oeschger, 'Perspectives on Global Change Science', *Quarternary Science Reviews* 19 (2000), pp. 37-44, here pp. 38-9.

48. National Research Council, *Changing Climate: reports of the CO2 assessment committee* (Washington: NAS, 1983); Stephen Seidel and Dale Keys, *Can We Delay a Greenhouse Warming?* (Washington: EPA, 1983); Hecht and Tirpak, 'Framework agreement', here p. 380; Goldemberg et al, *Energy for a sustainable world*, pp. 45–6.
49. 'Statement by the UNEP/WMO/ICSU International conference', in B. Bolin, B. Doos, J. Jager and R Warrick (eds.), *The Greenhouse Effect, Climatic Change and Ecosystems* (Chichester: SCOPE/John Wiley, 1986), pp. xx–xxiv; Bert Bolin, *A History of the Science and Politics of Climate Change: the role of the IPCC* (Cambridge: Cambridge University Press, 2007), pp. 38–9.
50. World Commission on Environment and Development, *Our Common Future* (WCED: Oslo, 1987), paragraphs 23–5.
51. Bolin, *A History of the Science*, pp. 44–9; Bodansky, 'Prologue', pp. 48–9.
52. Author's calculation from *BP Statistical Review of World Energy*.

Chapter 9

1. Author's calculation, from *BP Statistical Review*.
2. IPCC, 'Overview', in *Climate Change: The 1990 and 1992 IPCC Assessments* (Geneva: WMO and UNEP, 1992), pp. 52–7.
3. William Nitze, 'A failure of presidential leadership', in Mintzer and Leonard, *Negotiating Climate Change*, pp. 187–200; Hecht and Tirpak, 'Framework agreement', pp. 371–3.
4. WRI, *World Resources 1990–91*, especially pp. 13–20, 24–30 and 101–111; Anil Agarwal and Sunita Narain, *Global Warming in an Unequal World: a case of environmental colonialism* (New Delhi: Centre for Science and Environment, 1991).
5. UNFCCC, Articles 2 and 3; J. Goldemberg, 'The road to Rio', in Mintzer and Leonard, *Negotiating Climate Change*, pp. 175–85.
6. IPCC, *Second Assessment: Climate Change 1995 Report* (Geneva: IPCC, 1995), pp. 1–18; Ross Gelbspan, *The Heat is On: the high stakes battle over Earth's threatened climate* (Reading, Mass: Addison-Wesley, 1997), pp. 86–90 and 102–113; Robert Manne, 'How Vested Interests Defeated Climate Science', *The Monthly*, August 2012; Geoffrey Supran and Naomi Oreskes, 'Assessing ExxonMobil's climate change communications', *Environmental Research Letters* 12 (2017) 084019.
7. Gelbspan, *The Heat is On*, pp. 86 and 123–5; WRI, *World Resources 1998–99* (New York: Oxford University Press, 1998), pp. 174–5; David Held, Eva-Maria Nag and Charles Roger, *The Governance of Climate Change in China* (London: LSE, 2011), pp. 34–5 and 39.
8. UNFCCC, Article 2.
9. Angus Maddison, *The World Economy*, pp. 140–41; Ursula Huws, *Labor*, pp. 20–21; Joseph Stiglitz, *The Roaring Nineties: a new history of the world's most prosperous decade* (New York: W.W. Norton & co, 2003); Robin Broad and John Cavanagh, 'The Death of the Washington Consensus?' in W. Bello, N. Bullard and K. Malhotra (eds.), *Global Finance: new thinking on regulating speculative capital markets* (London: Zed, 2000), pp. 83–95; McNally, 'From financial crisis'.
10. See Ramachandra Guha, *Environmentalism: a global history* (New York: Longman, 2000), pp. 98–124, and Joan Martinez-Alier, *The Environmentalism of the Poor: a study of ecological conflicts and valuation* (Cheltenham: Edward Elgar, 2002), especially pp. 100–152.
11. Michael Pollitt, 'The role of policy in energy transitions: lessons from the energy liberalisation era', *Energy Policy* 50 (2012), pp. 128–37.
12. IEA, *Taxing Energy: why and how* (Paris: IEA, 1993), pp. 141 and 152–3. The research covered Denmark, Germany, Australia, Japan and the USA.

13. Larsen and Shah, *World Fossil Fuel Subsidies*; IEA, *World Energy Outlook 1999*, p. 64; Doug Koplow, *Measuring Energy Subsidies Using the Price-Gap Method: What Does It Leave Out?* (Winnipeg: IISD, 2009).
14. Doug Koplow, 'Global Energy Subsidies', in Antoine Halff, Benjamin K. Sovacool, and Jon Rozhon (eds.), *Energy Poverty: Global Challenges and Local Solutions* (Oxford: Oxford University Press, 2014), pp. 316–37; World Bank, *Subsidies in the Energy Sector: An Overview* (Washington: World Bank, 2010), pp. 47–62; UNEP, *Reforming Energy Subsidies: opportunities to contribute to the climate change agenda* (Nairobi: UNEP, 2008), pp. 12–14.
15. Lawrence Summers, 'The Case for Corrective Taxation', *National Tax Journal* 44:3 (September 1991), pp. 289–92; Eric Pooley, *The Climate War: true believers, power brokers and the fight to save the earth* (New York: Hyperion, 2010), pp. 85–9; Janet Milne, 'Carbon Taxes in the US: the context for the future', in J. Milne et al., *The Reality of Carbon Taxes in the 21st Century* (Vermont: Vermont Law School, 2008), pp. 1–30.
16. UNFCCC, Article 4, paragraphs 1(a)-(c); IEA, *Worldwide Trends*, pp. 27, 31, 36, 39 and 77; IEA, *Tracking Clean Energy Progress*, p. 7.
17. FERC web site, www.ferc.gov; Wollman et al, 'From public service'; Richard Rosen, Marjorie Kelly and John Stutz, *A Failed Experiment: why electricity deregulation did not work and could not work* (Boston: Tellus Institute, 2007).
18. Patterson, *Transforming Electricity*, pp. 14–15.
19. Hausman et al, *Global electrification*, pp. 265–9; Steve Thomas, 'Viewpoint: The Seven Brothers', *Energy Policy* 31 (2003), pp. 393–403; Martin, *In the Public Interest?*, pp. 108–111.
20. World Bank, *The World Bank's Role in the Electric Power Sector* (Washington: World Bank, 1993), pp. 11–18 (quoted); K. N. Gratwick and A. Eberhard, 'Demise of the standard model', *Energy Policy* 36 (2008), pp. 3948–60; Williams and Ghanadan, 'Electricity reform', *Energy* 31 (2006), pp. 815–44.
21. Dubash and Rajan, 'India: Electricity Reform'; Kale, *Electrifying India*, pp. 50–57.
22. Williams and Ghanadan, 'Electricity reform'.
23. Dubash and Rajan, 'India: Electricity Reform'; Pachauri, *An Energy Analysis*, pp. 34–6.
24. Brendan Martin, *Privatisation Pinoy-style* (London: Public World, 2002); Anupama Sen et al, *Reforming Electricity Reforms? Empirical evidence from Asian economies* (Oxford: OIES, 2016), p. 6; Karekezi and A. Sihag, *"Energy Access" Working Group Report*, pp. 8–9; Hall, *Electrifying Africa*, pp. 4–5.
25. D. Victor and T. Heller, 'Introduction' in Victor and Heller (eds.), *The Political Economy*; Anton Eberhard et al., *Underpowered: the state of the power sector in sub-Saharan Africa. Summary of background paper 6, Africa Infrastructure Country Diagnostic* (World Bank, June 2008), pp. 2–3.
26. Dubash, 'Introduction', in Dubash (ed.), *Power Politics*; Sen et al, *Reforming Electricity Reforms?*, pp. 38–9.
27. Navroz Dubash, 'The changing global context for power reform' in Dubash (ed.), *Power Politics: equity and environment in electricity reform* (Washington: World Resources Institute, 2002), pp. 11–30, and 'Conclusion', in the same volume, pp. 157–71. The research covered Argentina, India, Indonesia, Bulgaria, Ghana and South Africa.
28. Maddison, *The World Economy*, p. 149; McNally, 'From financial crisis', pp. 62–3; Stiglitz, *The Roaring Nineties*, pp. 217–18.
29. IEA, *Energy Use in the New Millennium* (Paris: IEA, 2007), pp. 30–31; IEA, *Thirty Years*, p. 41; Stevens, 'Oil markets', pp. 125–6; Yergin, *The Quest*, pp. 87–108.
30. Williams and Ghanadan, 'Electricity reform'; Frances Seymour and Agus Sari, 'Indonesia: electricity reform under economic crisis' in Dubash (ed.), *Power Politics*, pp. 75–96; Steve Thomas, 'The grin of the Cheshire cat', *Energy Policy* 34:15 (2006), pp. 1974–83.

31. Jonathan Sinton and David Fridley, 'What goes up: recent trends in China's energy consumption', *Energy Policy* 28 (2000), pp. 671–87; IEA, *World Energy Outlook* 1999, pp. 104–108; Xunpeng Shi, 'Can China's coal industry be reconciled with the environment?', in Ligang Song and Wing Thye Woo (eds.), *China's Dilemma* (Washington: Brookings Institution Press, 2008), pp. 367–91.
32. IEA, *World Energy Outlook* 1999, pp. 148–50; Sanjeev Gupta, Marijn Verhoeven, Robert Gillingham, Christian Schiller, Ali Mansoor, Juan Pablo Cordoba, *Equity and Efficiency in the Reform of Price Subsidies: A Guide for Policymakers* (Washington: IMF, 2000).
33. E. P. Thompson, 'The Moral Economy of the English Crowd in the Eighteenth Century', *Past and Present* 50 (1971), pp. 76–136.
34. McDonald, 'Electric Capitalism', van Heusden, 'Discipline', and Greg Ruiters, 'Free basic electricity in South Africa: a strategy for helping or containing the poor?', pp. 248–63, all in D. McDonald (ed.), *Electric Capitalism*.
35. Prishani Naidoo and Ahmed Veriava, 'From local to global (and back again?): anti-commodification struggles of the Soweto Electricity Crisis Committee', in McDonald (ed.), *Electric Capitalism*, pp. 321–37; Bond and Ngwane, 'Community Resistance'.
36. Simon Pirani, *Elusive Potential*, pp. 27–8; S. Pirani, *Change in Putin's Russia* (London: Pluto, 2010), pp. 39–41, 43–4 and 173–5.
37. 'Stealing Infrastructure Access in Brazil', *Brazil Business*, 30 April 2012; 'Power Theft Spurs Demand for Smart Meters at Brazilian Utilities', Bloomberg [news agency], 9 March 2012; USAID, *Transforming Electricity Consumers into Customers: Case Study of a Slum Electrification and Loss Reduction Project in Sao Paulo, Brazil* (Washington: USAID, 2009); USAID, *Innovative Approaches to Slum Electrification* (Washington: USAID, 2004).
38. Chhavi Dhingra, Shikha Gandhi, Akanksha Chaurey and P. K. Agarwal, 'Access to clean energy services for the urban and peri-urban poor: a case-study of Delhi', *Energy for Sustainable Development* 12:4 (2008), pp. 49–55.
39. Lamia Zaki, 'Transforming the City from Below: shantytown dwellers and the fight for electricity in Casablanca', in: Stephanie Cronin (ed.), *Subalterns and Social Protest: history from below in the Middle East and North Africa* (Abingdon: Routledge, 2008), pp. 116–37.
40. 'Power to the powerless', *The Economist*, 27 February 2016; Michael Greenstone, *Energy, Growth and Development* (London: IGC, March 2014).
41. For a different, thought-provoking, view, see: David Victor, *Global Warming Gridlock: creating more effective strategies for protecting the planet* (Cambridge: Cambridge University Press, 2011), especially pp. 203–240.
42. See 'Yes, I'd lie to you', *Economist*, 10 September 2016.
43. This phrase is often attributed to President George H. W. Bush I, but I have not found any original sources confirming that. It was, though, repeated as a mantra by US representatives at Rio. 'Rich vs. poor', *Time*, 1 June 1992; Raneet Panjabi, *The Earth Summit at Rio: politics, economics and the environment* (Boston: Northeastern University Press, 1997), p. 240; John Vidal, "Rio+20. Earth summit dawns", *Guardian*, 19 June 2012.
44. This point is made in Allwood et al., 'Industry 1.61803'.
45. See Achim Brunnengraber, 'The political economy of the Kyoto Protocol', *Socialist Register* 2007, pp. 213–30, and A. Brunnengraber, 'Kyoto's "flexible mechanisms" and the right to pollute the air', *Critical Currents* 7 (2009), pp. 26–35.

Chapter 10

1. IEA, *World Energy Investment Outlook* 2014 (Paris: IEA, 2014), p. 52; IEA, *Redrawing the Energy-Climate Map: WEO Special Report* (Paris: IEA, 2013), p. 15.

2. IEA, *Redrawing*, pp. 32-3.
3. Chateau and Lapillonne, *Energy demand*; IEA, *Worldwide Trends*, p. 43.
4. Heinz Schandl et al, 'Global Material Flows and Resource Productivity: forty years of evidence', *Journal of Industrial Ecology* (June 2017); Thomas Wiedemann et al, 'The material footprint of nations', *PNAS* 112:20, 19 May 2015, pp. 6271-6; Smil, *Making the Modern World*, especially pp. 119-56.
5. Stiglitz, *The Roaring Nineties*, pp. 5-6; David McNally, 'From financial crisis', pp. 63-5.
6. Author's calculations from *BP Statistical Review*. In the decade to 2000 commercial energy consumption rose 14.5 per cent (16.3 per cent in the OECD and 12.3 per cent outside it); in the decade to 2010, it rose 28.2 per cent (3.6 per cent in the OECD and 60.7 per cent outside it).
7. Zhang Tao et al, 'Analysis of National Coal-Mining Accident Data in China, 2001-2008', *Public Health Reports* 126 (March-April 2011), pp. 270-75; IEA, *World Energy Investment Outlook 2014*, p. 53; IEA, *Redrawing*, p. 9; Johansson et al, *Global Energy Assessment*, pp. 1143-4; David Robinson and Xin Li, *Closing Coal in China: international experiences to inform power sector reform* (Oxford: Smith School of Enterprise and the Environment, 2017), p. 16; 'Hot and bothered: special report on climate change', *Economist*, 28 November 2015.
8. Allwood, Cullen et al, *Sustainable Materials*, pp. 89-91; ABB, *The state*, pp. 11-12; L. Bernstein and J. Roy (eds.), 'Industry', in *Climate Change 2007: Mitigation. Contribution of Working Group III to the Fourth Assessment Report of the IPCC* (Cambridge: Cambridge University Press, 2007), pp. 739-810; IEA, *Tracking Clean Energy Progress*, p. 34; Johansson et al, *Global Energy Assessment*, p. 516.
9. Yergin, *The Quest*, pp. 163-5 and 181-4; Alvin Lin, Fuqiang Yang and Jason Portner, 'Global Energy Policy: a view from China', in A. Goldthau (ed.), *The Handbook of Global Energy Policy* (Chichester: Wiley Blackwell, 2013), pp. 391-406; C. Allsopp and B. Fattouh, 'The oil market: context, selected features and implications' in A. Goldthau (ed.), *The Handbook*, pp. 81-97; IEA, *World Economic Outlook 2014*, p. 49; Berck and Roberts, 'History of Prices'.
10. Marianne Haug, 'Clean energy and international oil', *Oxford Review of Economic Policy* 27:1 (2011), pp. 92-116, here p. 95.
11. Masami Kojima, *Petroleum Product Pricing and Complementary Policies: experiences of 65 developng countries since 2009* (Washington: World Bank, 2013), pp 4-5. Trevor Alleyne et al, 'Reforming Energy Subsidies: lessons from experience', pp. 23-41, here p. 34, Antonio David et al, 'Case Studies from the sub-Saharan Africa region', pp. 43-74, here p. 50, and Masahiro Nozaki and Baoping Shang, 'Case studies from emerging and developing Asia', pp. 75-86, all in B. Clements et al (eds.), *Energy Subsidy Reform: lessons and implications* (Washington: IMF, 2013).
12. Doug Guthrie, *China and Globalization: the social, economic and political transformation of Chinese society* (London: Routledge, 2012), pp. 95-6; Garnaut et al, 'China's rapid emissions growth'.
13. Nathaniel Aden, Nina Zheng and David Fridley, 'How can China lighten up? Urbanization, industrialization and energy demand scenarios', in B. Childers and M. Downing (eds.), *Energy Consumption in China* (New York: Nova Science, 2012), pp. 1-96, here pp. 2-3.
14. Aden, Zheng and Fridley, 'How can China Lighten Up?', pp. 14-21; Alvin Lin et al, 'Global Energy Policy'.
15. The proportions of embodied emissions were calculated for 2004. Ming Xu, Ran Li, John Crittenden and Yongshen Chen, 'CO_2 emissions embodied in China's exports from 2002 to 2008: a structural decomposition analysis', *Energy Policy* 39 (2011), pp. 7381-8; Davis and Caldeira, 'Consumption-based accounting'.
16. Aden, Zheng and Fridley, 'How can China Lighten Up?', here p. 1; Pachauri and Jiang, *The Household Energy Transition*, pp. 7-9; Nan Zhou, Michael McNeil and Mark

Levine, 'Energy for 500 million homes: drivers and outlook for residential energy consumption in China', in B. Childers and M. Downing (eds.), *Energy Consumption in China*, pp. 97–122, here pp. 106–7.
17. Newman and Kenworthy, *The End of Automobile Dependence*, p. 80; Smil, *China's Past*, p. 26; Yergin, *The Quest*, p. 221.
18. Day-ahead reserve prices (i.e. prices in contracts for delivery of electricity the following day, the dominant trading instrument) rose 120-fold. The aggregate electricity bill for all customers in California was $7.25 billion in 1999 and $33.5 billion in 2000 – 'the greatest interstate wealth transfer in US history' as Amory Lovins, *Small is Profitable*, pp. 57–60, put it.
19. James Sweeney, *The California Electricity Crisis* (Stanford: Hoover Institution Press, 2002), pp. 1–2 and 125–9; John Jurewitz, 'California's Electricity Debacle: a Guided Tour', *The Electricity Journal*, May 2002, pp. 10–29; Severin Borenstein et al, 'Measuring Market Inefficiencies in California's Restructured Wholesale Energy Market', *American Economic Review* 92:5 (2002), pp. 1376–1405; R. Rosen et al, *A Failed Experiment*.
20. Olga Gore et al, 'Russian electricity market reform: deregulation or reregulation?', *Energy Policy* 41 (2012), pp. 676–85; IEA, *Russian Electricity Reform 2013 Update* (Paris: IEA, 2013); S. Pirani, 'Sources of demand in the Russian domestic gas market', in J. Henderson and S. Pirani (eds.), *The Russian Gas Matrix: how markets are driving change* (Oxford: Oxford University Press, 2014), pp. 155–80.
21. IEA, *World Energy Investment Outlook 2014*, p. 92; David Buchan and Malcolm Keay, *Europe's Long Energy Journey: towards an energy union?* (Oxford: Oxford University Press, 2015), pp. 27–8; Rolf Wustenhagen and Emanuela Menichetti, 'The influence of energy policy on strategic choices for renewable energy investment' in A. Goldthau (ed.), *The Handbook of Global Energy Policy* (2013), pp. 373–87, here p. 374; 'Utilities wage campaign against rooftop solar', *Washington Post*, 7 March 2015.
22. Johannson et al, *Global Energy Assessment* (2012), p. 685; Dholakia, *Technology and Consumption*, pp. 101–5; David Nye, 'Consumption of Energy'; Sanchez et al, 'Miscellaneous electricity in US homes: historical decomposition and future trends', *Energy Policy* 26:8 (1998), pp. 585–93; Smil, *Energy in Nature and Society*, p. 267; US EIA, *Annual Energy Review* 2011, p. 41 and p. 53.
23. Data centre estimates by the *New York Times* ('Power, pollution and the internet', 22 September 2012) and Greenpeace (*Make IT Green: Cloud Computing and its Contribution to Climate Change* (Amsterdam: Greenpeace International, 2010)). Broader estimates by Greenpeace (*How Clean is Your Cloud* (Amsterdam: Greenpeace International, 2012) and Mark Mills of Digital Power Group (see 'The surprisingly large energy footprint of the digital economy', *Ecocentric*, 14 August 2013). See also Hughes, *The Bleeding Edge*, pp. 210–49.
24. Quintin Schiermeier, 'The Kyoto Protocol: hot air', *Nature*, 28 November 2012; Michael Le Page, 'Was Kyoto climate deal a success?', *New Scientist*, 14 June 2016; Igor Shishlov, Romain Morel and Valentin Bellassen, 'Compliance of the parties to the Kyoto Protocol in the first commitment period', *Climate Policy* 16:6 (2016), pp. 768–82.
25. Climate Focus, *The Paris Agreement: Summary* (December 2015); Heiner Flassbeck, 'The Paris Climate Change Agreement is an Epic Failure', *Flassbeck-Economics*, www.flassbeck-economics.com, 7 August 2016; Jamie Morgan, 'Paris COP 21: power that speaks the truth?', *Globalizations* 13 (2016), pp. 943–51.
26. D. Helm, 'EU Climate-change policy – a critique', in D. Helm and C. Hepburn (eds.), *The Economics and Politics of Climate Change* (Oxford: Oxford University Press, 2009), pp. 222–44; Buchan and Keay, *Europe's Long Energy Journey*, pp. 10–15.
27. Europa web site, https://ec.europa.eu/clima/policies/ets_en; IEA, *Redrawing*, p. 9, p. 16 and pp. 22–3; Brunnengraber, 'Kyoto's "flexible mechanisms"'; M. Pollitt, 'The role of policy in energy transitions', p. 134.
28. IEA, *Tracking Clean Energy Progress*, p. 22.

29. Margaret Oge, *Driving*, p. 65; Held et al, *The Governance*, pp. 29–30; Lin et al, 'Global Energy Policy', pp. 391–3; Sylvie Cornot-Gandolphe, *China's Coal Market: can Beijing tame 'King Coal'?* (Oxford: OIES, 2014), pp. 9–12.
30. Lee Schipper, *Automobile Fuel Economy and CO2 Emissions in Industrialized Countries: Troubling Trends through 2005/6* (Washington: WRI, 2006); the Pew Charitable Trusts, *History of Fuel Economy: one decade of innovation, two decades of inaction* (Washington: Pew Charitable Trusts, 2007); Oge, *Driving*, pp. 65–9; Johansson et al, *Global Energy Assessment*, p. 584.
31. Andre Serrenho, Jonathan Norman and Julian Allwood, 'The impact of reducing car weight on global emissions', *Phil Trans R Soc* (2017) A375:20160364, state that car manufacturers could have, as of 2016, sold 200 mpg models; current fuel efficiencies are mostly 20–30 mpg. See also: IEA, *Tracking Clean Energy Progress*, p. 42; Allwood et al., 'Industry 1.61803'.
32. Kimura, 'The role of standards'; H. Wilhite, E. Shove, Loren Lutzenhiser and W. Kempton, 'The legacy of twenty years of energy demand management: we know more about individual behaviour but next to nothing about demand', in E. Jochem, J. Sathaye and D. Bouille (eds.), *Society, Behaviour and Climate Change Mitigation* (Dordrecht: Kluwer, 2000).
33. IEA, *Tracking Clean Energy Progress* 2015, p. 46; Johannson et al, *Global Energy Assessment*, pp. 1156–61.
34. The Global Subsidies Initiative estimate of $400 billion/year in fossil fuel subsidies, including $100 billion/year in producer subsidies, published in 2010, was widely cited (GSI, *Relative Subsidies to Energy Sources: GSI Estimates* (Geneva: GSI/IISD, 2010)). A figure of 'more than $300 billion/year' probably undercounts the full expense, David Victor wrote. (Victor, *Untold Billions*, p. 9.) The World Bank, IEA, OPEC and OECD in 2010 estimated total energy subsidies, mostly to fossil fuels, at more than $700 billion/year. (IEA, OPEC, OECD, World Bank, *Analysis of the Scope of Energy Subsidies and Suggestions for the G-20 Initiative* (Toronto, 2010), pp. 4–5.) In 2012 the IEA published an estimate of $523 billion/year of total subsidies. (IEA, *World Energy Outlook 2012*, p. 23). See also Robert Bacon, Eduardo Ley and Masami Kojima, *Subsidies in the Energy Sector: an overview* (Washington: World Bank, 2010). These estimates were made using the price-gap method. (See p. 142.)
35. OECD, *Inventory of Estimated Budgetary Support and Tax Expenditures for Fossil Fuels* (Paris: OECD, 2012), pp. 27–38; OECD, *Companion to the Inventory of Support Measures for Fossil Fuels 2015* (Paris: OECD, 2015), especially pp. 9–10. The six countries added were Brazil, China, India, Indonesia, Russia and South Africa.
36. Michael Keen, Ian Parry and Jon Strand, *Planes, Ships and Taxes: charging for international aviation and maritime emissions* (Dublin: Economic Policy Panel Meeting, 2013); IMF, *Market-Based Instruments for International Aviation and Shipping as a Source of Climate Finance* (Washington: IMF, 2011).
37. Elizabeth Bast et al, *The Fossil Fuel Bailout: G20 subsidies for oil, gas and coal exploration* (Washington: ODI/Oil Change International, 2014); Doug Koplow, *Subsidies to Energy Industries* (Cambridge, Mass.: Earth Track, 2015), p. 7; D. Koplow, *Phasing out Fossil Fuel Subsidies in the G20: a progress update* (Washington: Earthtrack/Oil Change International, 2012).
38. Author's calculations from *BP Statistical Review* and World Bank data.
39. Ed Davey, UK climate change secretary, said 'these figures show green growth is achievable'. See 'Coal bust is driving drop in emissions', *New Scientist*, 21 March 2015.
40. The *BP Statistical Review* showed coal consumption falling by 1–2 per cent/year in 2014 and 2015. On Chinese statistics, see: Dabo Guan et al, 'The gigatonne gap in China's Carbon Dioxide Inventories', *Nature Climate Change* 2 (2012), pp. 672–5; Bing Xue and Wanxia Ren, 'China's uncertain CO_2 emissions', *Nature Climate Change* 2 (2012), p. 762; Zhu Li et al, 'Steps to China's carbon peak', *Nature* 522 (2015), pp. 279–81.

41. IEA, *Redrawing*, p. 9; IEA, *World Energy Investment Outlook* 2016, p. 29; Sylvie Cornot-Gandolphe, *The role of coal in South East Asia's power sector and implications for global and regional coal trade* (Oxford: OIES, 2016), pp. 1–2.
42. IEA, *World Energy Investment Outlook* 2016, p. 29; Christine Shearer et al, *Boom and Bust 2017: tracking the global coal plant pipeline* (San Francisco: Coalswarm, March 2017), p. 3.
43. IEA, *World Energy Outlook* 2014, p. 208; IEA, *Electricty Information* 2017, p. vii. The 'renewables' category covers geothermals, solar, wind and tidal power.
44. IEA, *World Energy Investment Outlook* 2016, p. 15, pp. 21–3 and p. 106; IEA, *World Energy Investment Outlook* 2014, pp. 20–21; IEA, *WEO 2015 Special Report - Energy and Climate*, p. 17.
45. Johansson et al, *Global Energy Assessment*, pp. 586–7; Newman and Kenworthy, *The End of Automobile Dependence*, pp. 22–7 and p. 103; UN Habitat, *Planning and Design for Sustainable Urban Mobility. Global Report on Human settlements* 2013 (Nairobi: UN Habitat, 2013), p. 21 and p. 35; David Metz, 'Peak Car and Beyond: the fourth era of travel', *Transport Reviews* 33 (2013), pp. 255–70; IEA, *Tracking Clean Energy Progress* 2015, p. 40.

Chapter 11

1. Bonneuil and Fressoz, *The Shock*, especially pp. 72–9 and 170–97 and 209–215; J. Donald Hughes, *An Environmental History of the World* (London: Routledge, 2001), especially pp. 52–79 and pp. 133–41.
2. Rachel Carson, *Silent Spring* (London: Hamish Hamilton, 1963); Barry Commoner, *The Closing Circle: confronting the environmental crisis* (London: Cape, 1972); Murray Bookchin, *Ecology and revolutionary thought* (New York: Times Change, 1970); M. Bookchin, *The ecology of freedom: the emergency and dissolution of hierarchy* (Edinburgh: AK Press, 2005); Ian Angus, 'Barry Commoner and the Great Acceleration', www.climateandcapitalism.com, 29 June 2014.
3. UN World Commission on Environment and Development, *Our Common Future* [the Brundtland commission report, 1987], paragraph 27. A 1980 report from the WWF, UNEP and International Union for the Conservation of Nature used the term 'sustainable development', but the Brundtland commission report became much better known. See also: Francis Sandbach, 'The rise and fall of the limits to growth debate', *Social Studies of Science* 8:4 (1978), pp. 495–520.
4. Jeffrey Sachs, *The Age of Sustainable Development* (New York: Columbia University Press, 2015), p. 12; Desta Mebratu, 'Sustainability and sustainable development'; Bonneuil and Fressoz, *The Shock*, p. 55 (citing IUCN); Donella Meadows, Jorgen Randers and Dennis Meadows, *Limits to growth: the 30-year update* (Vermont, Chelsea Green, 2004), pp. 238–63. A critique of 'ecosystem services' is: Sara Nelson, 'Beyond the Limits to Growth debate: ecology and the neoliberal counter-revolution', *Antipode* 47:2 (2015), pp 461–80.
5. Neva Goodwin et al, *Microeconomics in Context* (New York, M.E. Sharpe, 2009), pp. 251–2; Adam Smith, *An Inquiry into the Nature and Causes of the Wealth of Nations* [1776] (New York: Wm Benton, 1952), p. 287; James Otteson, *Adam Smith* (New York: Continuum books, 2011), pp. 93–6; Joseph Schumpeter, *History of Economic Analysis* [1954] (London: Routledge, 1997), pp. 576–7.
6. Karl Marx, *Capital*, vol. 1 [1867] (London: Lawrence & Wishart, 1977).
7. Lewis Mumford, *Technics and Civilisation*, p. 153; R. H. Tawney, *The Acquisitive Society* (London, G. Bell & sons, 1924), p. 33.
8. J.K. Galbraith, *The Affluent Society* (London: Penguin, 1998), p. 117, pp. 126–7.

9. On consumerism, see N. Goodwin, F. Ackerman and D. Korn (eds.), *The Consumer Society* (Washington: Island Press, 1997); Alan Durning, *How Much is Enough?: the consumer society and the future of the earth* (London: Norton, 1992); Teodor Scitovsky, *The Joyless Economy: the psychology of human satisfaction* (New York: Oxford University Press, 1992); Ben Fine and Ellen Leopold, *The World of Consumption* (London: Routledge, 1993). On needs and scarcity, see Nicholas Xenos, *Scarcity and Modernity* (London: Routledge, 1989) and Lyla Mehta (ed.), *The Limits to Scarcity: contesting the politics of allocation* (London: Earthscan, 2010).
10. Allan Schnaiberg, *The Environment: from surplus to scarcity* (Oxford: Oxford University Press, 1980), pp. 167-90. Later, in the 2000s, Schnaiberg, with colleagues, analysed US capitalism in the post-war period, and characterised the rising capital intensity and energy intensity of production processes as a 'treadmill of production'. Kenneth Gould, David Pellow and A. Schnaiberg, *The treadmill of production: injustice and unsustainability in the global economy* (Boulder, Co: Paradigm, 2008). For a critique of the 'treadmill of production' idea by other Marxist writers, see *Organisation & Environment* 17:3 (September 2004), including Erik Olin Wright, 'Interrogating the treadmill of production', pp. 317-22 and Richard York, 'The treadmill of (diversifying) production', pp. 355-62.
11. Thomas Malthus, *Essay on the Principle of Population* [1798-1826] (London: Penguin Books, 2015); Ronald Meek (ed.), *Marx and Engels on Malthus* (London: Lawrence & Wishart, 1953); Yves Charbit, *Economic, Social and Demographic Thought in the XIXth Century: the population debate from Malthus to Marx* (Dordrecht: Springer, 2009), especially pp. 1-42, 121-8 and 146-56.
12. Garrett Hardin, 'The Tragedy of the Commons', *Science* 162 (1968), pp. 1243-8; Eric Ross, *The Malthus Factor: poverty, politics and population in capitalist development* (London: Zed Books, 1988).
13. Atiq Rahman and Annie Roncerel, 'A view from the ground up', in Irving Mintzer and J. Amber Leonard, *Negotiating Climate Change* (1994), pp. 239-73, here p. 263; Atiq Rahman, Nick Robins and Annie Roncerel (eds.), *Consumption versus population: which is the climate bomb?* (Dhaka: University Press, 1998).
14. Paul Ehrlich, Peter Kareiva and Gretchen Daily, 'Securing natural capital and expanding equity to rescale civilization', *Nature*, 7 June 2012, pp. 68-73.
15. IPCC, *Climate Change 2014. Mitigation of Climate Change (Working Group 3 contribution to the Fifth Assessment Report)*, p. 47.
16. Thomas Piketty, *Capital in the twenty-first century* (London: Belknap Press, 2015); Branko Milanovic, *Global Inequality: a new approach for the age of globalization* (Cambridge, Mass: Harvard, 2016).
17. Fritz Schumacher, *Small is Beautiful: a study of economics as if people mattered* [1973] (London: Vintage, 2011); Herman Daly, *Beyond Growth: the economics of sustainable development* (Boston: Beacon Press, 1996).
18. Nicholas Stern, *The Economics of Climate Change: the Stern Review* (Cambridge: Cambridge University Press, 2008). See also: Ernst von Weizsacker, A. Lovins and L. Lovins, *Factor Four: doubling wealth, halving resource use* (London: Earthscan, 1998). Critiques of 'green growth' include: Gareth Dale, Manu Mathai and J.A. Puppim de Olivera (eds.), *Green Growth: ideology, political economy and the alternatives* (London: Zed Books, 2016); Daniel Tanuro, *Green Capitalism: why it can't work* (London: Merlin, 2013).
19. See Andre Gorz, *Ecology as Politics* (London: Pluto Press, 1987). Relevant recent literature includes Jason Moore, *Capitalism in the Web of Life: ecology and the accumulation of capital* (London: Verso, 2015); and John Bellamy Foster, Brett Clark and Richard York, *The Ecological Rift: capitalism's war on the earth* (New York: Monthly Review Press, 2010). A presentation for a general readership, focussed on global warming, is: Naomi Klein, *This Changes Everything: capitalism vs the climate* (London: Allen Lane, 2014).

Chapter 12

1. The research resulted in a major publication: Grubler and Wilson (eds.), *Energy Technology Innovation*. For the points cited, see pp. 5–6.
2. A. Grubler and C. Wilson, 'Lessons from the history of technological change for clean energy scenarios and policies', *Natural Resources Forum* 35 (2011), pp. 165–84.
3. See also Rosenberg, *Exploring the Black Box*. On 'technological lock-in', see G. Unruh, 'Understanding carbon lock-in', *Energy Policy* 28 (2000), pp. 817–30; Unruh, 'Globalising carbon lock-in', *Energy Policy* 34 (2006), pp. 1185–97; Arnulf Grubler, Nebojsa Nakicenovic and David Victor, 'Dynamics of energy technologies and global change', *Energy Policy* 27 (1999), pp. 247–80. On the closely related idea of 'path dependency', see: Hughes, *Networks of Power*; Paul David, 'Clio and the Economics of QWERTY', *American Economic Review* 75:2 (1985), pp. 332–7.
4. Sovacool, 'How Long Will it Take?'
5. *Energy Policy* 50 (2012) is a special issue devoted to the proceedings. Here I cite the editorial, 'Past and prospective energy transitions: Insights from history', pp. 1–7.
6. Mark Jacobson et al, '100% Clean and Renewable Wind, Water and Sunlight All-Sector Energy Roadmaps for 139 Countries', *Joule* 1 (2017), pp. 108–21; M. Jacobson et al, 'Providing all global energy with wind, water and solar power' (parts 1 and 2), *Energy Policy* 39 (2011), pp. 1154–90. See also Greenpeace International et al, *Energy [R]evolution: a sustainable world energy outlook 2015* (Amsterdam: Greenpeace International, 2015).
7. K. Anderson and A. Bows, 'Reframing the climate challenge in light of post-2000 emissions trends', *Phil Trans R Soc* (2008) 2008.0138. On the second world war, see Bill McKibben, 'A World at War', *The New Republic*, 15 August 2016; Laurence Delina and Mark Diesendorf, 'Is wartime mobilisation a suitable policy model for rapid national climate mitigation?', *Energy Policy* 58 (2013), pp. 371–80. Klein, *This Changes Everything*, pp. 16–17, pp. 115–16 and pp. 452–6, engages critically with the analogy of wartime 'collective sacrifice', and argues that a range of twentieth-century labour and civil rights movements are more relevant. See also Malm, *Fossil Capital*, pp. 384–8.
8. IEA *Electricity Information* 2017, p. vii. Renewables contribute 4.9 per cent of global electricity.
9. 'A world turned upside down', *Economist*, 25 February 2017; 'The green revolution is stalling', *New Scientist*, 5 August 2017; Malcolm Keay, David Rhys and David Robinson, 'Electricity markets and pricing for the distributed generation era' in F. Sioshansi (ed.), *Distributed Generation and its Implications for the Utility Industry* (Amsterdam: Academic Press, 2014), pp. 165–87.
10. Heinberg and Fridley, *Our Renewable Future*, pp. 47–80; Pedro Prieto and Charles Hall, *Spain's Photovoltaic Revolution: the energy return on investment* (New York: Springer, 2012); Megan Guilford et al, 'A new long-term assessment of EROI for US oil and gas', *Sustainability* 2011, pp. 1866–87; 'The real EROI of photovoltaic systems: Professor Hall weighs in', www.resilience.org, 27 May 2016. See also: F.P. Sioshansi (ed.), *Smart Grid: integrating renewable, distributed and efficient energy* (Oxford: Elsevier, 2012); Sioshansi (ed.), *Distributed Generation*.
11. http://wardsauto.com/news-analysis/world-vehicle-population-tops-1-billion-units.
12. James Arbib and Tony Seba, *Rethinking Transportation 2020-2030* (Washington: RethinkX, May 2017); Heinberg and Fridley, *Our Renewable Future*, p. 84.
13. Heinberg and Fridley, *Our Renewable Future*, pp. 95–102 and 106–112.
14. 'Mark Carney warns investors face "huge" climate change losses', *Financial Times*, 29 September 2015.
15. 'Fossil fuel burning set to hit record high in 2017', *Guardian*, 13 November 2017; Corinne Le Quere et al, 'Global Carbon Budget 2017', *Earth Systems Science Data Discussions*, November 2017.

16. A cautionary note was sounded in: Trade Unions for Energy Democracy, *Energy Transition: Are We Winning?* (New York: TUED, 2017).
17. For example, IEA, *Energy Technology Perspectives* (Paris: IEA, various years); Allwood et al, *With Both Eyes Open*; David Mackay, *Sustainable Energy - Without the Hot Air* (Cambridge: UIT, 2009); Weizsacker et al, *Factor Four*.
18. I have written about this in detail elsewhere, e.g. the conclusions to S. Pirani, *The Russian Revolution in Retreat 1920-24* (London: Routledge, 2008).
19. Paul Mason, *Post Capitalism: a guide to our future* (London: Allen Lane, 2015); Nick Srnicek and Alex Williams, *Inventing the Future: postcapitalism and a world without work* (London: Verso, 2015), and '#Accelerate Manifesto for an accelerationist politics', *Critical Legal Thinking*, 14 May 2013.
20. Philip Bedall and Achim Brunnengraber, '"Participation NGOs" and "protest NGOs" in international climate politics', in *Global Transformations Towards a Low Carbon Society*, Global Transformations Towards a Low Carbon Society Working Paper series no. 10 (Hamburg: University of Hamburg, 2014), pp. 44–62. See also: Elmar Altvater and Achim Brunnengraber (eds.), *After Cancun: climate governance or climate conflicts* (Berlin: VS Research, 2011).
21. Here the authors refer to German Advisory Council on Global Change, *World in Transition – a social contract for sustainability* (Berlin: WBGU, 2011); UNEP, *Towards a green economy: pathways to sustainable development and poverty eradication* (Nairobi: UNEP, 2011); New Economics Foundation, *A Green New Deal: joined-up policies to solve the triple crunch of the credit crisis, climate change and high oil prices* (London: NEF, 2008); Tim Jackson, *Prosperity Without Growth: economics for a finite planet* (London: Earthscan, 2009); Niko Paech, *Liberation from Excess: the road to a post-growth economy* (Munich: Oekom Verlag, 2012).
22. Studies on current impacts include: World Bank, *Turn Down the Heat: climate extremes, regional impacts and the case for resilience* (Washington: World Bank, 2013); Yohannes GebreMichael et al, *More than climate change: pressures leading to innovation by pastoralists in Ethiopia and Niger* (Brighton: Future Agricultures Consortium, 2011); New Economics Foundation, *Counting on Uncertainty: the economic case for community-based adaption in North East Kenya* (London: NEF, 2012).
23. Rebecca Solnit, *A Paradise Built in Hell: the extraordinary communities that arise in disaster* (London: Viking, 2009).

Appendix 1

1. Barry Commoner, *Making Peace With the Planet* (The New Press, New York, 1990), p. 150.
2. Paul Ehrlich, John Holdren and Barry Commoner, 'A Bulletin dialogue on The Closing Circle', *Bulletin of the Atomic Scientists* 1972 (vol. 28, no. 5), pp. 16–17 and 42–56; B. Commoner, 'The environmental cost of economic growth', pp. 339–63, and P. Ehrlich and J. Holdren, 'Impact of population growth', pp. 365–77, in R. Ridker (ed.), *Population, Resources and the Environment. The US Commission on Population Growth and the American Future research reports*, vol. III (Washington: Government Printing Office, 1972).
3. Marian Chertow, 'The IPAT equation and its variants: changing views of technology and environmental impact', *Journal of Industrial Ecology* 4:4, 2001, pp. 13–29. Unfortunately the author of an intellectual history of Ehrlich's work, and a biographer of Commoner, both focus on the intemperate form of their debate rather than the substance. See Paul Sabin, *The Bet: Paul Ehrlich, Julian Simon and our gamble over Earth's future* (London: Yale University Press, 2013) and Michael Egan, *Barry*

Commoner and the science of survival: the remaking of American environmentalism (London: MIT Press, 2007).
4. Nick Robins, 'How Much is Sustainable?', in Atiq Rahman, Nick Robins and Annie Roncerel (eds.), *Consumption versus population: which is the climate bomb?* (Dhaka: University Press, 1998), pp. 1-8.
5. James Bruce, Hoesung Lee and Erik Haites (eds.), *Climate Change 1995. Economic and social dimensions of climate change. Contribution of the Working Group III to the Second Assessment Report of the IPCC* (IPCC/Cambridge UP, 1995), p. 27.
6. The key sections of the IPCC's Fifth Assessment Report of 2014 on the causes of greenhouse gas emissions rely heavily on work by researchers using the Kaya Identity. See: IPCC, *Climate Change 2014: Mitigation of Climate Change. Contribution of Working Group III to the Fifth Assessment Report of the IPCC* (Cambridge: Cambridge University Press, 2014), pp. 125-37 ('Historical, current and future trends') and pp. 364-85 ('Key drivers of global change'). The sentence quoted is from the 'Summary for Policymakers', the only part of the report most non-specialists read, p. 8.
7. Thomas Dietz and Eugene Rosa, 'Rethinking the environmental impacts of population, affluence and technology', *Human Ecology Review* 1 (1994), pp. 277-300; Donella Meadows quoted in M. Maniates, 'Individualization: plant a tree, buy a bike, save the world?', in Princen et al, *Confronting Consumption*, pp. 43-66.
8. See Thomas Dietz and Andrew Jorgenson (eds.), *Structural Human Ecology: new essays in risk, energy and sustainability* (Washington: Washington State University Press, 2013), especially chapters 1, 8 and 10.
9. E. Rosa and T. Dietz, 'Human drivers of national greenhouse-gas emissions', *Nature Climate Change* 2 (2012), pp. 581-6; Arthur Mol and Gert Spaargaren, 'Towards a Sociology of Environmental Flows', in Gert Spaargaren, Arthur Mol and Frederick Buttel (eds.), *Governing Environmental Flows: global challenges to social theory* (London: MIT Press, 2006), pp. 39-82.
10. Ehrlich, Holdren and Commoner, 'A Bulletin dialogue'; Bruce Hannon, 'Bottles, Cans, Energy', *Environment* 14:2 (1972), pp. 11-21; A. Makhijani and A. Lichtenberg, 'Energy and well-being', *Environment* 14:5 (1972), pp. 10-18.
11. Howard Odum, *Environment, Power and Society* (New York: John Wiley & Sons, 1971); Odum, 'Energy, Ecology and Economics', *Ambio* 2:6 (1973), pp. 220-27; Spreng, *Net-Energy Analysis*, pages 125 and 136-8.
12. Anne Carter, 'Application of Input-Output Analysis to Energy problems', *Science*, 19 April 1974, pp. 325-9; Christian Estrup, 'Energy Consumption Analysis by Application of National Input-Output Tables', *Industrial Marketing Management* 3 (1974), pp. 193-210.
13. Clark Bullard and Robert Herendeen, 'The energy cost of goods and services', *Energy Policy* December 1975, pp. 268-78; Clark Bullard, Peter Penner and David Pilati, 'Net energy analysis: handbook for combining process and input-output analysis', *Resources and Energy* 1 (1978) pp. 267-313; David A. Huettner, 'Net energy analysis: an economic assessment', *Science* no. 4235 9 April 1976, pp. 101-4; Marina Fischer-Kowalski and Helmut Haberl, 'Tons, Joules and Money: Modes of Production and their Sustainability Problems', *Society & Natural Resources* 10 (1997), pp. 61-85; Luc Gagnon, 'Civilisation and energy payback', *Energy Policy* 36 (2008), pp. 3317-22.
14. Spreng, *Net-Energy Analysis*; Murphy, 'The implications of declining energy return'; Robert Herendeen, 'Net energy considerations', in R.E. West and F. Kreith (eds.), *Economic Analysis of Solar Thermal Energy Systems* (London: MIT Press, 1988), pp. 255-73.
15. Robert Ayres and Leslie Ayres, *Industrial ecology: closing the materials cycle* (Cheltenham: Edward Elgar, 1996); Mol and Spaargaren, 'Towards a Sociology of Environmental Flows'; Wilhite et al, 'The legacy of twenty years'.
16. For example, Allwood, Cullen et al, *Sustainable Materials*.

17. Goldblatt, *Sustainable energy consumption*; Goldblatt, 'A dynamic structuration'; R. Dholakia and N. Dholakia, 'From social psychology to political economy: a model of energy use behaviour', *Journal of Economic Psychology* (1983) 3, pp. 231–47.
18. A major quantitative research project is reported in: Michael Raupach et al, 'Global and regional drivers of accelerating CO_2 emissions', *PNAS* 104 (2007), pp. 10288–93.
19. Davis and Caldeira, 'Consumption-based accounting'.
20. Shoibal Chakravarty et al, 'Sharing global CO_2 emission reductions among one billion high emitters', *Proceedings of the National Academy of Sciences of the USA* 106:29 (2009), pp. 11884–8; A. Grubler and S. Pachauri, 'Problems with burden-sharing proposal among one billion high emitters', *Proceedings of the National Academy of Sciences of the USA* 106:43 (2009), pp. E122–E123.
21. Chancel and Piketty, *Carbon inequality*.
22. Edgar Hertwich and Glen Peters, 'Carbon Footprint of Nations: a global, trade-linked analysis', *Environmental Science & Technology* 43:16 (2009), pp. 6414–20.
23. BP: www.bp.com/en/global/corporate/energy-economics/statistical-review-of-world-energy.html. The Shift Project: www.tsp-data-portal.org/.
24. IEA, *Energy Statistics Manual*, pp. 136–7; Robert Wilson, 'Do Renewables Lower Energy Consumption?', *The Energy Collective*, 18 February 2014 www.theenergycollective.com/robertwilson190/338991/does-renewable-energy-lower-energy-intensity; H. Douglas Lightfoot, 'Understand the three different scales for measuring primary energy and avoid errors', *Energy* 32 (2007), pp. 1478–83.

Further reading and bibliography

All the material consulted is referred to in the notes to the chapters. It is also listed in a complete Bibliography, posted on line at <http://bit.ly/222Read>. The books and articles that proved most important for my research are listed below.

General works about or partly about fossil fuel consumption. For the whole period since 1950: Vaclav Smil, *Energy in World History* (Oxford: Westview Press, 1994), and *Energy in Nature and Society* (London: MIT Press, 2008). **For 1950–80:** Joel Darmstadter, *Energy in the World Economy* (Baltimore: Johns Hopkins Press, 1971); Arjun Makhijani and Alan Poole, *Energy and Agriculture in the Third World* (Cambridge, Mass: Bellinger, 1975); Erik Eckholm, *The Other Energy Crisis: Firewood* (Washington: Worldwatch Institute, 1975); Darmstadter et al, *How Industrial Societies Use Energy* (Baltimore: Johns Hopkins Press, 1977); Joy Dunkerley, *Trends in Energy Use in Industrial Societies* (Washington: RFF, 1980); Vaclav Smil and William Knowland (eds.), *Energy in the Developing World* (Oxford: Oxford University Press, 1980); Bertrand Chateau and Bruno Lapillonne, *Energy Demand: facts and trends* (Vienna: Springer, 1982); John Clark, *The Political Economy of World Energy* (Hemel Hempstead: Harvester, 1990). **From the 1980s onwards:** Jose Goldemberg et al, *Energy for a sustainable world* (New York: Wiley, 1988); UNDP, *World Energy Assessment* (2000); IEA, *Thirty Years of Energy Use in IEA Countries* (2004), *Energy Use in the New Millennium* (2007), and other publications; the IPCC, *Climate Change 2014: Mitigation* and other publications; Thomas Johannson et al, *Global Energy Assessment* (Cambridge: IIASA/ Cambridge University Press, 2012).

On energy flows through technological systems: Daniel Spreng, *Net Energy Systems* (New York: Praeger, 1988); Julian Alwood, Jonathan Cullen et al, *Sustainable Materials With Both Eyes Open* (Cambridge: UIT, 2012); Arnalf Grubler and Charlie Wilson (eds.), *Energy Technology Innovation: learning from historical successes and failures* (Cambridge: Cambridge University Press, 2014); Allwood et al., 'Industry 1.61803: the transition to an industry with reduced material demand fit for a low carbon future', *Philosophical Transactions of the Royal Society A* vol. 375, no. 2095 (13 June 2017) p. 20160361. **On electricity networks and electricity reform:** Richard Rudolf and Scott Ridley, *Power Struggle: the hundred-year war over electricity* (New York: Harper & Row, 1986); Walt Patterson, *Transforming Electricity* (London: RIIA, 1999); Navroz Dubash (ed.), *Power Politics: equity and environment in electricity reform* (Washington: WRI, 2002); J. Williams and R. Ghanadan, 'Electricity reform in developing and transition countries: a reappraisal', *Energy* 31 (2006), pp. 815–44; David Victor and Thomas Heller (eds.), *The Political Economy of Power Sector Reform* (Cambridge: Cambridge University Press, 2007).

On final consumers: Paul Stern and Elliott Aronson (Eds.), *Energy Use: the human dimension* (New York: Freeman, 1984); David Goldblatt, *Sustainable Energy Consumption and Society* (Dordrecht: Springer, 2005); Juliet Schor, 'Prices and Quantities', *Ecological Economics* 55 (2005), pp. 309–20; Ruby Dholakia, *Technology and Consumption* (New York: Springer, 2012). **On final consumers in developing countries:** Elizabeth Cecelski et al, *Household Energy and the Poor in the Third World* (Washington: RFF, 1979); Shonali Pachauri and Leiwen Jiang, *The Household Energy Transition in India and China* (Laxenburg: IIASA, 2008). **On methodological approaches to consumption:** Allan Schnaiberg, *The Environment: from surplus to scarcity* (Oxford: Oxford University Press,

1980); Marina Fischer-Kowalski and Helmut Haberl, 'Tons, Joules and Money', *Society & Natural Resources* 10 (1995), pp. 61–85; Neva Goodwin et al (eds.), *The Consumer Society* (Washington: Island Press, 1997); Daniel Spreng et al (eds.), *Tackling Long Term Energy Problems* (Dordrecht: Springer, 2012).

On the discovery of global warming and international climate agreements: Irving Mintzer and J. Leonard (eds.), *Negotiating Climate Change: the inside story of the Rio convention* (Cambridge: Cambridge University Press, 1994); Alan Hecht and Dennis Tirpak, 'Framework agreement on climate change: a scientific and policy history', *Climatic Change* 29 (1995), pp. 371–402; Spencer Weart, *The Discovery of Global Warming* (London: Harvard University Press, 2003); Bert Bolin, *A History of the Science and Politics of Climate Change* (Cambridge: Cambridge University Press, 2007); Achim Brunnengraber, 'The Political Economy of the Kyoto Protocol', *Socialist Register* 2007, pp. 213–30.

On fossil fuel subsidies: Gerard Brannon, *Energy Taxes and Subsidies* (Cambridge, Mass: Ballinger, 1974); the World Bank, Subsidies in the Energy Sector (2010); OECD, *Inventory of Estimated Budgetary Support and Tax Expenditures for Fossil Fuels 2013* (2012); Doug Koplow, *Phasing out Fossil Fuel Subsidies in the G20* (Washington: Earthtrack, 2012), and *Subsidies to Energy Industries* (Cambridge, Mass: Earth Track, 2015).

On the USA: David Nye, *Electrifying America* (London: MIT Press, 1990) and *Consuming Power* (London: MIT Press, 2001); Stan Luger, *Corporate Power, American Democracy and the Automobile Industry* (Cambridge: Cambridge University Press, 2000); Matthew Huber, *Lifeblood: oil, freedom and the forces of capital* (Minneapolis: Quadrant, 2013).
On China: Vaclav Smil, *China's Past, China's Future* (London: Routledge, 2004); Jonathan Sinton and David Fridley, 'What goes up: recent trends in China's energy consumption', *Energy Policy* 28 (2000), pp. 671–87; Brian Childers and Margret Downing (eds.), *Energy Consumption in China* (New York: Nova Science, 2012). **On India:** Shonali Pachauri, *An Energy Analysis of Household Consumption* (Dordrecht: Springer, 2004), and 'Household electricity access a trivial contributor to CO2 emissions growth', *Nature Climate Change* 4 (2014), pp. 1073–6; Sunila Kale, *Electrifying India* (Stanford: Stanford University Press, 2014). **On South Africa:** David McDonald (ed.), *Electric Capitalism* (London: Earthscan, 2009). **On the Soviet Union:** Leslie Dienes and Theodore Shabad, *The Soviet Energy System* (Washington: Winston, 1979); Jonathan Coopersmith, *The Electrification of Russia* (Ithaca: Cornell University Press, 1992).

For the historiographical context for this book, and relevant publications, see the Introduction. For material on the history of fossil fuel consumption prior to 1950, see the notes to Chapter 1. For material on statistical methodologies, see Appendix 1.

Index

AES 118, 144
Africa and African countries 88, 93, 105, 109, 129, 131, 146, 148, 158
agriculture 32, 72-3, 88-90, 104, 115-16, 129
  'green revolution' in 32, 72, 89, 115
  industrial 41, 72-3, 79, 82, 190
  see also fertilisers, chemical
air conditioning 20, 33, 74, 86, 130, 165-6
Allen, Robert 12
aluminium 32, 34, 41, 67, 126, 158, 159
America, Central 104
Anderson, Kevin 184
Anthropocene epoch 1, 9-10, 212
appliances, electrical 20, 33, 74-5, 86-7, 90-2, 109-10, 120, 126-7, 130-1, 155, 160, 194
  see also radio, refrigerators, television, washing machines
'appropriate technology' movement 103
Argentina 93, 96
Asia and Asian countries 88, 123, 140, 146, 147-8, 155, 158, 168
Australia 86, 142, 166
aviation 23, 31, 70, 71, 166
Ayres, Robert 28

Bahrain 52
Bangladesh 104
Baran, Paul 84
Bebel, August 39
Bedall, Philip 191
Belgium 82
biofuels 4, 40, 74, 88-9, 105-6, 113-14
biomass *see biofuels*
Belgium 18
Bonneuil, Christophe 3, 135, 173
Bookchin, Murray 173
Bows, Alice 184
BP (British Petroleum) 139-40, 147
BP Statistical Review 56, 206-7
Brazil 52, 93, 96, 104, 105, 108, 150, 158, 169

Brundtland commission 137, 173-4, 204
Brunnengraber, Achim 191
buildings 33, 46, 73-4, 166, 187, 194
  see also air conditioning, cities, construction and heat
Bush, George H.W. (US president 1989-93) 138
Bush, George (US president 2001-2009) 165

Canada 81, 99, 123, 166
  in climate negotiations 140
carbon budget, global 57-8, 181
Carbon Majors Database 53-4
Carbon Tracker Initiative 58
cars 21, 31, 43, 44-5, 63, 71, 79, 99, 104, 127-8, 129, 130, 155, 160, 168-9, 194
  electric 31, 186
  fuel efficiency of 31, 84, 101, 102, 128, 165
  manufacture of 21, 33, 68, 82, 84, 124
  manufacturers 21-2, 102, 131
  ownership rates 44-5, 71, 83-5, 88, 129, 160, 168
  style changes 23, 84
  see also sport utility vehicles
Carson, Rachel 173
Carter, Jimmy (US president) 35, 103, 131
cement 32, 41, 67-8, 82, 100, 129, 159
Chancel, Lucien 51
Chateau, Bertrand 85-6
chemicals, manufacture of 67-9, 100
Chevron 147
Chicago Convention (1944) 23
Chile 93, 132
China 47, 56, 69, 86, 90, 106, 122, 123, 127, 129, 130, 134-5, 142, 148, 154-8, 159-60, 168, 180, 183, 187-8, 195
  coal use in eleventh century 11
  electricity in 52, 112-14
  energy balances of 62-5
  in climate negotiations 139, 165
  wind power in 35

Chrysler 21, 84, 102, 139
cities 13, 43–4, 75, 85, 88, 148–9, 160, 189, 219
   infrastructure 44–5, 189
   lighting 13
   poverty in 40, 44, 89, 148–51
   transport in 22, 84
climate change *see global warming*
climate negotiations, international 1, 151–2, 162–5, 182, 185, 191, 196, 205
   Toronto (1988) 137
   Rio summit (1992) 137, 138–40, 143, 151, 177–8
   Kyoto (1997) 140, 151, 163
   Copenhagen (2009) 1, 163
   Paris (2015) 2, 51, 57, 164, 188
   see also Intergovernmental Panel on Climate Change
climate science *see global warming*
climate science denial, 131, 139
Clinton, Bill (US president) 139, 142
coal
   consumption of 18, 39, 54, 79, 114, 124, 127, 153, 155, 157–8, 165, 167–8, 187–8
   and Industrial Revolution 11–13
   in late nineteenth century 14
   manufacturers 139
   price of 134
   production of 14, 28, 141, 157, 168
   trade of 19, 38
cogeneration 29–30, 103, 126, 132, 183, 188
colonialism *see imperialism*
combined-cycle gas turbines (CCGT) 29, 31, 124, 143
combined heat and power (CHP) *see cogeneration*
commercial and public services 67
Commoner, Barry 173, 177, 201–2, 203–4
computers and information technology 36–7, 123, 126, 155, 162, 166, 197
construction 32, 34, 68, 73–4, 82, 187
cooking and cookers 33, 74, 86, 131, 160
consumerism 45, 75, 130, 174–6, 193
consumption (general) 45–6, 174–6, 177–8, 193
consumption-based accounting 53, 159–60, 205–6
Coopersmith, Jonathan 111

countryside *see rural areas*
Cowan, Ruth Schwartz 90
Czechoslovakia 129

Daly, Herman 179
debts, international 104
deforestation 105–6, 131, 139
'dematerialisation' 125, 155
Denmark 16, 30, 35, 102, 162, 168, 183, 185
   in climate negotiations 139
Devine, Warren 27
diesel engine 17, 18
Dietz, Thomas 202–3
Dubash, Navroz 116, 145–6

ecological economics 179–80
economic growth 104, 167, 173–4, 178–80, 182, 186, 187–8, 191, 197
economic theories 149, 151, 173–6, 178–80
economy, world 20–1, 95, 122–4, 140–1, 155–8, 167, 179, 194–5
Edison, Thomas 14, 16
Egypt 93
Ehrlich, Paul 177–8, 201–2, 203–4
electric-arc furnace (EAF) 32, 68, 126
electricity 14–15, 17, 20, 33, 66–7, 90, 127, 129, 155, 160–2, 184–5, 210
   access, and lack of access, to 40, 42, 52, 107–10, 114–16, 117–18, 119–20, 148–51, 183, 195
   as a commodity 20, 38–40, 103, 119–20, 194
   autonomous production of 113–14, 115, 117
   households' ability to pay for 109–10, 120, 148–51
   market reforms of (1980s-1990s) 109, 115–16, 118, 132–3, 140–1, 143–7
   production of 18, 26, 29, 36–7, 61, 81, 82, 99, 101, 132, 143, 157, 165
   supplied as a public service 16, 20, 38–40, 119–20, 194
electricity networks 15–16, 30–1, 36–7, 61, 81, 103, 106, 111–12, 113–14, 194
   decentralised (i.e. distributed generation) 31, 37, 39, 184–6, 188, 189

electrification 20, 21, 41, 81, 89–90, 103, 107–21, 130, 191, 194–5
  rural 20, 21, 108–10, 112
energy
  commercial system of 4, 19–20, 40–1
  commodification of 4, 38, 148–51, 193
  consumption of 19, 61–2, 81, 97, 122, 208 *see also fossil fuel consumption*
  definitions of and forms of 4, 26, 58–64
  non-commercial forms of 4, 20, 40–1, 88–9, 105–6
  substituted for labour *see labour, energy substituted for*
  supplied as a state benefit 4, 38–40
  renewable 4, 5, 34–6, 134, 153, 161–2, 165, 168, 185–6, 196, 207 *see also wind, solar power*
energy accounting 26, 53–4, 159–60, 206–7 *see also consumption-based accounting and net energy analysis*
energy balances 5, 18, 59–66 *see also energy, consumption of*
energy conservation 26, 94, 99, 101–3, 125–6, 129, 143, 166, 168, 188–9
'energy crisis' (1970s) 93–4
energy efficiency *see energy conservation*
energy return on energy invested (EROI) 58
energy services 26, 27–8, 182
energy systems 25–8, 181–4
Enron 116, 118, 144, 145, 160–1
environmentalism 27, 30, 102, 124, 134, 173–4, 201
'environmentalism of the poor' 141
Environmental Protection Agency (EPA) (USA) 102, 136
ethanol 104
Europe and European countries 81, 83, 85, 87, 100, 143, 160, 162
  in climate negotiations 138–9, 152
  see also European Union
European Bank for Reconstruction and Development (EBRD) 129
European Union (EU) 69, 141, 161–2, 164–5
Exxon 118, 139
ExxonMobil 144, 147

Faraday, Michael 14

fertilisers, chemical 14, 24, 32, 41, 72, 83, 89–90, 158, 159
Filtzer, Donald 91
Finland 30
First World War 18
Ford 21, 84, 102, 139
Ford, Gerald (US president) 102
former Soviet countries 30, 149–50, 168
  in climate negotiations 140
fossil fuels
  prices of 28, 82, 96, 99, 101
  production of 28–31
  transformation of 4, 29–31, 60–2
  transition away from 181–90, 191–2, 196–7
  used as raw materials 4
  see also coal, gas and oil
fossil fuel consumption 41, 43–6, 122–3, 153–5, 156–7, 167–8, 193
  and population growth 47–50
  direct and indirect 45–6, 53, 187
  discretionary and non-discretionary 46, 176, 187, 205
  global overview 54–6, 59–60
  in rich countries 5, 54–6, 74–5, 82–7, 124–8, 155
  outside the rich world 5, 54–6, 74–5, 87–90, 128–31, 153–5
  per capita measurement of 20, 43, 50–2, 205–6
  statistical records of 19, 53–76, 201, 206–7
  technological means of reducing 33–4
France 18, 82, 83, 99, 100, 101, 183
Fressoz, Jean-Baptiste 3, 135, 173
Fridley, David 186
fuel efficiency *see cars, fuel efficiency of*

Galbraith, J.K. 175–6, 179
gas, natural 38, 157
  consumption of 18, 54–6, 74, 79, 124, 127, 132, 153, 167–8
  flaring of 58, 118, 231
  production of 28, 141, 168
  recovered from coal 13
  'unconventional' production of 28, 58, 157, 167
gas, liquefied natural (LNG) 29
General Electric 16, 23, 107

General Motors (GM) 21, 22, 84, 102
Gentle, Leonard 119
Germany 17, 18, 20, 23, 35, 51, 142, 162, 166, 168, 185–6
  East 100, 129
  West 99, 100, 122, 124
  in climate negotiations 139
Ghana 131
Global Energy Assessment (2012) 36–7, 41, 43, 52
global warming 1, 56–8, 135–7, 138–40, 146, 192, 195–6
  see also climate negotiations
globalisation 122–3, 140–1
gold production 119–20
Goldblatt, David 46, 127
Goldemberg, Jose 139
Gorz, Andre 180
greenhouse gas emissions 56–7, 159–60, 202–3, 205–6
Greenstone, Michael 151
Grubler, Arnalf 182

Haber-Bosch process 32
Hallwood, Paul 104
Hannon, Bruce 204
Harvey, David 132
Healy, Stephen 39
heat 66–7, 155
  for buildings (including households) 33, 73–4, 85–6, 127, 160
  for industry 69
  for water 86, 155
heat pumps 36
Heinberg, Richard 186
Heller, Thomas 146
Holdren, John 201–2
households 17, 45–6, 67, 74–5, 85–7, 89, 127, 130, 155, 160, 187
Hughes, Thomas 183
hydro power 10–11, 34, 113, 153

imperialism 13–14, 39
India 47, 83, 86, 89, 91, 93, 105, 122, 129, 130, 134, 142, 158, 168, 195
  electricity in 29, 47, 52, 108, 114–17, 144–5, 150
Indonesia 147, 148

Industrial Revolution 11–14, 16, 25, 182, 193
Industrial Revolution, second (late nineteenth century) 14–18, 25, 32, 182, 193
Industrial Revolution, third (late twentieth century), 25, 33, 36–7, 162, 183, 190, 196–7
industrialisation *see industry*
industry 11–13, 17, 33, 41, 62, 67–9, 82, 89, 100, 124, 126, 129, 158, 159, 187, 194, 211
  automation in 17, 43
  mass manufacturing techniques in 43, 75, 123, 127
  inequality 50–2, 64, 73, 74, 103, 130, 182–3, 205–6
Insull, Samuel 16, 21
Intergovernmental Panel on Climate Change (IPCC) 57, 137, 138–9, 202
internal combustion engine 14, 31 *see also cars*
International Energy Agency (IEA) 53, 59, 61, 96, 100, 124, 127, 142–3, 147, 164, 166, 168, 184, 206–7
International Institute for Applied Systems Analysis (IIASA) 103
International Monetary Fund (IMF) 105, 141, 148
international oil companies 80, 93–5, 124
Internet, the *see computers and information technology*
IPAT formula 202–3
Iran 80, 95, 142
iron *see steel and iron*
Israel 93–95
Italy 82, 99, 128
Iwayemi, Akin 118

Jacobs, Jane 84
Japan 35, 69, 71, 81, 82, 86, 96, 99–100, 101, 122, 126, 128, 157, 160, 165–6
Jensen, James 125, 126

Kale, Sunila 3, 115–16
Karbuz, Sohbet 70
Kaya identity 202–3
Kenworthy, Jeffrey 169
Kenya 145

Keynes, John Maynard 126
Keynesianism 132
Koplow, Doug 133
Korea, South 126, 147
Kropotkin, Petr 39
Kyoto protocol *see climate negotiations*

labour 20, 25–6, 41–3, 44, 123, 126–7, 159, 175, 179, 190, 194
   domestic 20, 40, 45–6, 74–5, 90–2, 190, 194
   energy substituted for 11, 12, 17, 34, 82, 126
labour movements 12, 123, 141, 180
labour process *see labour*
Landes, David 13
Lapillonne, Bruno 85–6
Lebergott, Stanley 92
Leontief, Wasily 204
Liberia 131
Libya 80, 94
Lieber, Robert 101
Limits to Growth report 102, 173–4, 203
Lovins, Amory 27–8, 33, 102–3, 161, 188

MacGill, Iain 39
Maddison, Angus 123
Makhijani, Arjun 89
Malm, Andreas 3, 13
Malthus, Thomas 177 *see also* neo-Malthusianism
Marx, Karl 174–5, 177
Marxism 2
Mason, Paul 190
materials manufacture 33–4, 41, 67, 79, 100, 155, 187
   see also aluminium, cement, 'dematerialisation', plastics, steel
McNally, David 123
Mexico 52, 89, 129
Middle East and Middle Eastern countries 94–5, 148
military, fuel consumption by 70
Mobil 118
Mol, Artur 203
Morgan, Jamie 163
Morocco 150–1
motor cars *see cars*
Mumford, Lewis 175

neo-Malthusianism 177–8
neoliberalism 132, 133, 140–2, 144, 152, 160–1
net energy analysis 103, 204, 206
Netherlands, the 20, 74, 82, 125, 127
Newman, Peter 169
Nigeria 64, 66, 106, 129, 131, 141, 148, 209
   electricity in 110, 117–19
Nixon, Richard (US president) 101–2
Nordhaus, William 103
nuclear power 34, 101, 124, 132, 134, 153, 183
Nye, David 3, 38–9

Obama, Barack (US president) 163, 165
'obsolescence, planned' 23
Odum, Howard 204
oil and oil products 14, 23, 38, 80–1, 89, 131
   consumption of 18, 54, 61, 74, 79, 94–6, 98, 99–100, 104–6, 130, 158
   prices of 80–1, 93–5, 96, 99–100, 104–6, 122, 127–8, 134, 147, 158, 166, 188, 195, 210
   production of 18, 28, 80, 93–5, 124, 141, 168, 210
   refining of 18, 61, 81
   trade of 19, 39, 81, 93–5, 124
   'unconventional' production of 28
   see also international oil companies
Olukoju, Ayodeji 117–18
Organisation for Economic Cooperation and Development (OECD) 5, 166–7, 168–9, 184, 212
Organisation of Petroleum Exporting Countries (OPEC) 93–5, 124
ozone layer 137

Pachauri, Shonali 51, 115
packaging 75
Pakistan 104
paper, manufacture of 67
Parsons, Charles 15
Patterson, Walt 31, 108, 143
Peru 110
petrochemicals 32, 61, 67–9, 126
petrol *see oil and oil products*
pets, domestic 73

Philippines, the 89, 104, 109, 145, 147
Piketty, Thomas 51
plastics 32, 67, 69, 75, 127
Podobnik, Bruce 13
Poland 122
Pollitt, Michael 141
pollution 13, 165, 168, 201
population growth 47–50, 114–15, 117, 177–8, 193, 201–2
post-war boom 33, 79–90
power generation *see electricity, production of*
public transport 22, 83–4, 88
Public Utility Regulatory Policies Act (PURPA) (USA) 103

Qatar 69

radio 20, 86–7, 130
railways 13, 17, 18, 71, 84, 194
Rajan, Sudhir 145
Reagan, Ronald (US president) 35, 128, 131–2, 204
refrigerators 20, 74, 83, 86–7, 127, 130, 160, 166
regulation, state 101, 123, 140, 144–7, 160–2, 165–6, 196
renewables *see energy, renewable*
residential consumption *see households*
Rio summit (1992) see *climate negotiations, international*
roads 22, 23, 44–5, 84, 104, 129, 135
Rosa, Eugene 202–3
rural areas outside the rich world 40, 88–90
  see also electrification, rural
Russia 18, 50, 51, 87, 142, 147, 149–50, 158, 161, 166
  in climate negotiations 164
  see also Soviet Union

Sachs, Jeffrey 174
Saro-Wiwa, Ken 118
Saudi Arabia 69, 80, 94
Schnaiberg, Allan 176
Schumacher, E.F. 179
Second World War 23–4, 70, 80
Shell 79, 118, 139–40, 147
shipping 18, 23, 71
  powered by steam 13, 17

Siemens-AEG 107
Sinclair, Stuart 104
slavery 11, 14
Smil, Vaclav 3, 50
Smith, Adam 174–5
social change 189–192, 197
social movements and social protests 124, 141, 148–50, 158, 191–2
socialism 39, 180, 184–5, 190
'soft energy paths' 28, 102
solar power 33, 34–5, 36–7, 134, 168
Solnit, Rebecca 192
South Africa 158
  electricity in 108, 119–20, 149
Sovacool, Benjamin 183–4
Soviet Union 23, 30, 56, 70, 81, 87–8, 91, 96, 122, 124, 180
  electricity in 21, 40, 87–8, 110–12
  in climate negotiations 138
  see also former Soviet countries and Russia
Spain 35, 45, 83, 162, 168, 185–6
Spaargaren, Gert 203
sports utility vehicles (SUVs) 128, 165
Spreng, Daniel 33, 126, 176, 204
Srnicek, Nick 190
steam engine 11–13
steel and iron 17, 32, 33–4, 61, 67–8, 82, 100, 126, 158, 159
Steinmetz, Charles 39
Stern Review, the (2007) 179, 184
Stevens, Paul 124
stranded assets 187–8
subsidies and taxes 46, 80, 99, 100, 101–2, 127–8, 133–5, 141–3, 158, 161–2, 166–7
suburbanisation 45, 83, 85
Sununu, John 138
sustainability 102, 173–4
Sweden 30, 36, 183
Sweezy, Paul 84
Switzerland 36, 75

Taiwan 69, 126
taxes *see subsidies and taxes*
Tawney, R.H. 175
technological revolution (late twentieth century) *see Industrial Revolution, third*

technologies and technological change 28–37, 173, 181–5, 188–90, 193–4
television 127, 130, 160
Texaco 140
Thailand 52, 109, 110, 145, 147
Thatcher, Margaret (UK prime minister) 132
Thompson, E.P. 149
Tooze, Adam 3
trade, international 71
  see also coal, trade of, and oil, trade of
transport 18, 20, 71, 168–9, 186
  see also aviation, cars, railways, roads, shipping
Trentmann, Frank 3
Trump, Donald (US president) 184, 192
turbine, gas 31
turbine, steam 15, 26
Turkey 93, 168

United Kingdom (UK) 75, 81, 83, 99–100, 123, 127, 128, 132–3, 166
  in the nineteenth century 12–13, 16
United Nations (UN) 102
United Nations Conference on Environment and Development (Rio, 1992) *see climate negotiations, international*
United Nations Development Programme (UNDP) 26, 125, 126
United Nations Environment Programme (UNEP) 136
United Nations Framework Convention on Climate Change (UNFCCC) *see climate negotiations, international*
United States of America (USA) 33, 35, 45, 50, 69, 70, 79–82, 83–5, 86–7, 90–1, 93–6, 99–100, 101–3, 122–4, 142, 143, 160, 166, 168, 187–8, 194, 204
  in climate negotiations 138–40, 151–2, 165

energy balances of 62–3
electricity and heat in 29, 30, 132, 160–1
  in the nineteenth century 13, 14, 16
  in early twentieth century 18, 21
  in Second World War 23–4
  military 70
urbanisation *see cities*
USSR *see Soviet Union*

Veblen, Thorstein 175
Venezuela 94, 106
Victor, David 146
Vietnam 45, 109
Volkswagen 165

washing machines 20, 74, 86–7, 127, 130, 131, 160
waste 69
water power *see hydro power*
Weart, Geoffrey 136
Weber, Max 175
Westinghouse 16, 107
Williams, Alex 190
Wilson, Charlie 182
wind power 11, 16, 35, 103, 168, 185–6
wood fuel 11, 40, 105–6
Workshop on Alternative Energy Strategies 103
World Bank 40, 107–8, 141, 142, 144–7, 184
World Commission on Environment and Development (1983) *see Bruntland commission*
World Energy Conference 103
World Meteorological Organisation 136
World Resources Institute 104, 125, 134, 139
World Trade Organisation 141
Wrigley, Edward 12

Zola, Emile 39–40